Construction technology 3

The technology of refurbishment and maintenance

MIKE RILEY

Director of the School of the Built Environment
Liverpool John Moores University, UK

and

ALISON COTGRAVE

Deputy Director of the
School of the Built Environment
Liverpool John Moores University, UK

Second edition

palgrave
macmillan

First edition 2005
This edition published 2011 by
PALGRAVE MACMILLAN

Palgrave Macmillan in the UK is an imprint of Macmillan Publishers Limited, registered in England, company number 785998, of Houndmills, Basingstoke, Hampshire RG21 6XS.

Palgrave Macmillan in the US is a division of St Martin's Press LLC, 175 Fifth Avenue, New York, NY 10010.

Palgrave Macmillan is the global academic imprint of the above companies and has companies and representatives throughout the world.

Palgrave® and Macmillan® are registered trademarks in the United States, the United Kingdom, Europe and other countries.

ISBN-13: 978–0–230–29014–3   paperback
ISBN-10: 0–230–29014–0   paperback

This book is printed on paper suitable for recycling and made from fully managed and sustained forest sources. Logging, pulping and manufacturing processes are expected to conform to the environmental regulations of the country of origin.

A catalogue record for this book is available from the British Library.

10 9 8 7 6 5 4 3 2
20 19 18 17 16 15 14 13 12 11

Printed and bound in Great Britain by
CPI Antony Rowe, Chippenham and Eastbourne

Contents

Preface   v
Acknowledgements   vii

PART 1   Background to the refurbishment and maintenance of buildings   1

1   The context of refurbishment   3
1.1 Definitions of refurbishment   5
1.2 The amount of refurbishment work undertaken in the UK and associated costs   8
1.3 Issues that affect the decision to refurbish   12
1.4 Refurbishment vs. redevelopment from an environmental perspective   18
1.5 Overview of statutory control of buildings   27

2   The context of maintenance   46
2.1 What is maintenance?   47
2.2 Building maintenance management   51

PART 2   Common defects encountered during construction   57

3   Common defects in buildings   59
3.1 Origins and mechanisms of defects   60
3.2 Analysis of defects   67
3.3 Substructure defects   69
3.4 Defects in walls, claddings and frames   72
3.5 Roof defects   87
3.6 Defects in non-timber floors   92
3.7 Timber defects   94
3.8 Dampness in walls   100
3.9 Flooding in buildings   106

PART 3   The technology of maintenance and refurbishment   113

4   Common refurbishment technologies   115
4.1 Underpinning   117
4.2 Waterproofing of basements   124
4.3 Façade retention   132
4.4 Overcladding   146

4.5	Overroofing and reroofing	156
4.6	Upgrading and retrofitting of building services	165
4.7	Remedying dampness	169
4.8	Repairs to masonry	173
4.9	Treatment of timber defects	176

PART 4   **Management of maintenance and refurbishment**   **179**

5	The management of refurbishment work	181
	5.1 Management of design	182
	5.2 Procurement and management of construction	186
6	Demolition and disposal	198
	6.1 The demolition decision	199
	6.2 Demolition techniques	202
7	Case studies	208
	7.1 Major refurbishment study	209
	7.2 'Decision to demolish' study	223
	Index	229

Preface

There are many texts available in the area of refurbishment and maintenance, resulting in an ever-expanding array of information for the student to manage. While such texts are invaluable reference sources, they are often difficult to use as learning vehicles.

This text differs in that it provides a truly student-centred approach to the technology of refurbishment and maintenance. Unlike reference texts that are used for selectively accessing specific items of information, this book is designed to be read as a continuous learning support. Uniquely, both housing *and* large-span, multi-storey commercial and industrial buildings are covered in detail within this one text, making this an ideal centre of learning from which the student can explore further. This book also looks at reuse and the environmental impact of refurbishing as well as new build, and so encourages the student to develop a good understanding of sustainability. In addition, content on the management of refurbishment contracts makes this text suitable for students of construction from both technology and management perspectives.

This volume builds on the subject matter that was introduced in the first two books in the series, *Construction Technology 1: House Construction, 2nd edition and Construction Technology 2: Industrial and Commercial Building, 2nd edition* although it is also a valuable standalone learning resource. This new edition increases the focus on the sustainability aspects of refurbishment and the use and occupation of buildings and provides further, detailed case studies to support the core material. The decision basis for demolition is considered in greater detail, together with the principles and practices of façade retention and these are supported by illustrated case studies.

A genuine learning text

The text is structured to provide a logical progression and development of knowledge, from basic introductory material to the more advanced concepts relating to the technology of refurbishment and maintenance. The content is aimed at students wishing to gain an understanding of the subject without the need to consult several different and costly volumes. Unlike reference texts that are used for selectively accessing specific items of information, this book is intended to be read as a continuous learning support vehicle. Whilst the student can access specific areas for reference, the major benefit can be gained by reading the book from the start, progressing through the various chapters to gain a holistic appreciation of the various aspects of refurbishment and maintenance.

Key learning features

The learning process is supported by several key features that make this text different from its competitors. These are embedded at strategic positions to enhance the learning process.

- Case studies include photographs and commentary on specific aspects of the technology of refurbishment and maintenance, enabling the student to visualise details and components in real situations.
- Reflective summaries are included to encourage the student to reflect on the subject matter and to assist in reinforcing the knowledge gained.
- Review tasks aid in allowing the reader to consider different aspects of the subject at key points in the text.
- The Info points incorporated into each chapter identify key texts or information sources to support the student's potential needs for extra information on particular topics.
- In addition, the margin notes are used to expand on certain details discussed in the main text without diverting the reader's attention from the core subject matter.

Website

A website supporting this book and designed to enhance the learning process can be found at http://www.palgrave.com/science/engineering/riley3. This contains outline answers to the reflective summaries and review tasks for each chapter.

MIKE RILEY
ALISON COTGRAVE

Acknowledgements

Paul Hodgkinson and Aseel Hussein for producing the diagrams.

Paul Murray, University of Plymouth, for allowing us to use material from the SLICE demolition learning packs.

Simon Hairsnape, Colin Davies and Richard Pinnington for providing the over-cladding case studies.

Peter Williams for his help and advice on health and safety.

Jack Rostron for his help relating to disabled access issues.

Laurie Brady for his assistance with proofreading.

Noora Kokkarinen for help with research.

Jules, Steve and Sam.

Acknowledgements

Background to the refurbishment and maintenance of buildings

The context of refurbishment

Aims

After studying this chapter you should be able to:

Explain the meaning of the terms *refurbishment, conversion, restoration, renovation* and *retrofit*

Appreciate the amount of refurbishment work that is undertaken in the UK, and the associated costs

Discuss the issues that will affect the decision to refurbish a building as opposed to demolition and new build

Show an understanding of the environmental implications of refurbishment as opposed to demolition and new build

Explain issues relating to the listing of buildings

Relate current Building Regulation requirements to refurbishment projects

Explain how health and safety legislation should be applied to the design and management of refurbishment contracts

This chapter contains the following sections:

1.1   Definitions of refurbishment
1.2   The amount of refurbishment work undertaken in the UK and associated costs
1.3   Issues that affect the decision to refurbish
1.4   Refurbishment vs. redevelopment from an environmental perspective
1.5   Overview of statutory control of buildings

Info point

- Building Regulations Approved Document L 2006 (2010 edition valid from 1st October 2010)
- Building Regulations Approved Document M 2004
- Chartered Institute of Building (CIOB) Construction Paper No. 66 1996: Characteristics and difficulties associated with refurbishment
- Building Research Establishment paper IP9/02 Part 1: Refurbishment or redevelopment of office buildings? Sustainability comparisons
- Building Research Establishment paper IP9/02 Part 2: Refurbishment or redevelopment of office buildings? Sustainability case histories
- CIRIA Report 113: A guide to the management of building refurbishment
- BRE Digest 446: Assessing environmental impacts of construction
- BRE Digest 452: Whole life costing and life cycle assessment for sustainable building design
- http://www.officescorer.info/
- http://envest2.bre.co.uk/-
- http://projects.bre.co.uk/refurb/nitecool/
- Edwards, B. (1999) *Sustainable Architecture: European Directives and Building Design*, 2nd edn, Architectural Press, Oxford, pp. xiv–xvi, 229
- Royal Institution of Chartered Surveyors (1986) *A Guide to Life Cycle Costing for Construction*, Surveyors Publications
- Webb, R. (2000) Sustainable architecture – cities, buildings and technology. Paper presented at Sustainable Building 2000 Conference, Maastricht

1.1 | Definitions of refurbishment

Introduction

- After studying this section you should be able to explain the differences between common terms associated with refurbishment work.
- You should also have developed an understanding of the variables that will affect the decision to refurbish a building and to what extent they will do so.
- In addition, you should be able to explain where refurbishment activity fits into the whole life cycle of a building.

Overview

There is a general acceptance that the decision to refurbish rather than to demolish and rebuild can have benefits in terms of sustainability and economics. The extent of refurbishment required will vary from situation to situation, but there are certain principles that can be applied in all cases. There is an escalating 'scale of intervention' which spans from superficial 'face lifting' of a building to extensive remodelling and upgrading. Between these two extremes the term 'refurbishment' may be applied to many different approaches in terms of technology utilised.

What is refurbishment?

Building work can be classified as either *new build* or *refurbishment*. New build is an easy concept to grasp, as it is a term applied to any work that is starting from scratch. There is no part of any structure left on a site. Refurbishment, however, is a more difficult concept to generalise.

A very broad definition of the term refurbishment is:

Work undertaken to an existing building

However, refurbishment schemes can take many forms and may be undertaken for a variety of very different reasons. There are also a number of terms that are commonly used to describe work undertaken to an existing building, and clarification of the exact meaning of these terms is essential. Throughout this book there will be references to these other terms, and the definitions used will be those of the authors. Terms used in other publications may be based on different definitions.

Refurbishment can be defined as:

Extending the useful life of existing buildings through the adaptation of their basic forms to provide a new or updated version of the original structure

The amount of work that is required in order to achieve this definition will be very different on different projects, and will depend on:

The value of buildings can be determined by assessing how fit for purpose and/or flexible they are.

- The condition of the existing structure
- The shape and size of the existing structure
- The location of the structure
- The intended use of the structure
- The amount of work required to the existing structure to enable compliance with current Building Regulations
- Whether the building is listed, either wholly or partly
- Adequate funding being available
- Whether the work can be carried out safely.

These issues will be discussed further in Section 1.4.

Other terms that are often used instead of or in conjunction with refurbishment are:

- *Conversion* implies that the main use of the building will be altered, but that the main structure will not be changed.
- *Renovation* and *restoration* imply that the work consists of renewal and repair only, and that the works carried out will simply address dilapidations to avoid further degradation of the building.
- *Retrofit* essentially means fitting new and more modern systems into an existing building. The term is commonly associated with building services because a common phenomenon in buildings is that the life of the building structure and fabric will be considerably longer than that of the installed services.

From these definitions it can be seen that refurbishment could include all of these elements on both a large or a small scale.

Figure 1.1
Life cycle of a building.

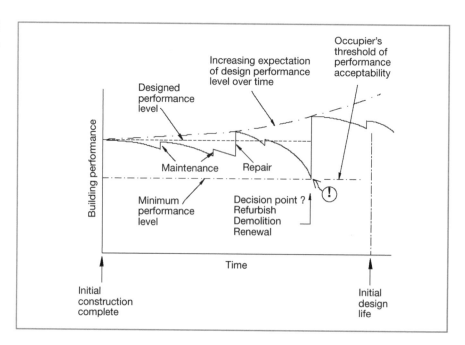

PART 1

The commonly used expression of barn conversion is an example of this confusion of terms. Certainly the barn is to be converted from a building that animals live in to a building that humans live in, and major works are required to create a suitable living arrangement. However, usually the external walls will remain, but will be restored and renovated. An entire services installation will be required, and this could be termed a retrofit, as a modern system is to be installed in an old building. In addition to this the useful life of the existing buildings will be extended through the adaptation of its basic form to provide a new version of the original structure, and therefore the building is being refurbished!

The term *refurbishment* can therefore be taken to mean that the existing building is not usable in its present form. Figure 1.1 illustrates where the refurbishment phase of a building fits into the whole life cycle of the building. The diagram does not show a specific time when refurbishment is required, as this may depend on the level of maintenance that has been undertaken during the occupation of the building; nor does it give actual values of performance requirement. A building could have been very well maintained but not meet the performance criteria of the existing or planned occupier, or very little maintenance may have been undertaken but refurbishment is not undertaken because the building owner/occupier has low performance requirements.

Owner/occupiers are expecting more from buildings than ever before, and the diagram illustrates how this trend towards increasing performance requirements will impact on the need for refurbishment.

Reflective summary

- The extent of refurbishment required will vary from situation to situation.
- There is an escalating 'scale of intervention' which spans from superficial 'face lifting' of a building to extensive remodelling and upgrading.
- *Refurbishment* can be defined as:

    Extending the useful life of existing buildings through the adaptation of their basic forms to provide a new or updated version of the original structure

- The amount of work that is required in order to achieve this definition will be very different on different projects and will depend on a number of factors.
- Terms that are often used instead of or in conjunction with refurbishment are *conversion*, *renovation*, *restoration* and *retrofit*.
- There is no 'fixed' time in the whole life of a building when refurbishment should take place. It will depend on the required building performance of the owner/occupier.

Review task

Explain the terms *refurbishment*, *conversion* and *retrofit*.

Discuss the issues that could affect the amount of work required in a refurbishment project.

1.2 | The amount of refurbishment work undertaken in the UK and associated costs

Introduction

- After studying this section you should have developed an understanding of the level of refurbishment work undertaken in the UK, and in which sectors of work refurbishment is most favourably viewed.
- You should be able to compare this with the amount of new build work undertaken in the various sectors.
- You should also be able to explain the problems associated with the costing of refurbishment work and what is meant by the term 'allowing for contingencies'.
- In addition you should be able to discuss the discrepancies that occur between tender price and actual completion costs for refurbishment work as opposed to new build.

Overview

Refurbishment work is undertaken extensively in the UK for a variety of reasons, such as buildings being of such merit that the replacement of the building is less desirable that its alteration and upgrading for future use (to be discussed in detail in Section 1.5) and most commonly because the building structure is sound, but the plan layout of the building is unsuitable for modern purposes. The owners of commercial buildings frequently specify that they require flexible floor plans that can easily be adapted to suit potentially different occupiers during the life of the building. Most older buildings have very rigid floor layouts, which will need extensive work carried out every time there is changed occupancy.

There is a common misconception that to refurbish will cost less than to demolish and build new, based on the idea that the amount of building work required will be reduced if most of the existing structure is to remain intact. However, this is not necessarily the case, and one of the main problems with refurbishment work is control of costs during the construction work.

The amount of refurbishment work undertaken

The value of all construction work excluding maintenance in the UK in 2000 was in the region of £60 billion. Of this, approximately half was classified as refurbishment work; in addition, nearly £50 billion is estimated to have been spent on maintenance. These figures demonstrate that refurbishment is a major element of the UK construction industry, and that trend is likely to grow during the next decade due to the regeneration of inner cities and towns, and the growing concern about building on greenfield sites. The environmental impact of refur-

Figure 1.2

Breakdown of construction
work undertaken in the UK.

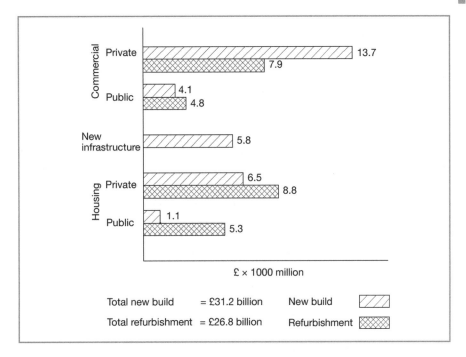

bishment work is less than that of new build work and this is another driver that will probably encourage clients to opt for refurbishment as opposed to new build.

A breakdown of the type of work undertaken that was classified as refurbishment compared to new build is shown in Figure 1.2.

If the figure stated for new infrastructure is removed from the overall figures, then refurbishment work actually accounts for more of the total value of works to buildings than does new build. The only area where new build work is still more popular than refurbishment is in the commercial private sector, and this could be accounted for by the problems associated with the costing of refurbishment work, and reduction of competitiveness of tenders for refurbishment work as opposed to new build.

Pricing of refurbishment work

One of the major problems with refurbishment work is the difficulty in determining a cost for the works before construction work starts. This is due to a number of factors:

■ There will always be a high level of 'unknowns', i.e. problems with the existing building that will only become apparent during construction work.
■ These items cannot be shown on a drawing, and therefore they are priced as the work proceeds. When work that is not shown on a drawing is required, the usual practice is to price it on a daywork basis, which will always be more expensive than bill rates.

Generally it is more difficult
to manage refurbishment
contracts when buildings
remain occupied.

- If the building to be refurbished is to remain occupied during the works, a large amount of money will need to be set aside to facilitate this.
- High levels of protection may be required to areas of the building that are listed, and it may be difficult to price this accurately before work starts.
- The work may be very 'bitty', with small amounts of work required all over the building. It is difficult to price this work accurately.
- Small amounts of materials may be required and it is difficult to get competitive prices for such small quantities.
- Health and safety issues may be more difficult to determine before the contract proceeds and may require additional funding while work is being undertaken.
- In any refurbishment projects some demolition is usually required. Restrictions may apply regarding noise control and the work may have to be undertaken out of hours, which will increase costs.
- In new build work, methods of work are fairly standard and can be priced accurately. This may not be the case in refurbishment work, and solutions to construction problems may need to be decided on-site. The solution to the problem may require extra funding above the amount allocated in the bill.
- Because of the difficulty in producing a Bill of Quantities, a great deal of refurbishment work is priced using drawings and a specification. The work to be undertaken will never be 'standard' and each tendering contractor will interpret the documents in a different way. This can lead to large differences in tender bids and can affect competition. In new build work very competitive prices can be achieved.
- It is common to come across what are now classified as dangerous materials in existing buildings (e.g. lead and asbestos). If the presence of these materials is not known at the time of the initial pricing, additional costs will be required in order to remove these materials. Specialist subcontractors are required for the removal of these materials and this can be an expensive process.
- While undertaking refurbishment, older technologies may be uncovered that need restoration and repair. This work could require specialist materials and labour and also the use of 'one-off' components, which will be more expensive than those that can be bought 'off the shelf'.
- All of the above could lead to the contract duration being increased. This will increase costs because, for example, site accommodation will be required for longer. From a client perspective this increased duration will mean that revenue cannot be generated from the building as early and potential rental income will be lost. If the proposed occupier is to move from another building into the newly refurbished building, there may have to be changes to the arrangements for the move, which could incur costs.

It is far more difficult to
produce accurate works
budgets for refurbishment as
opposed to new build work.

This potential for additional costs being incurred in refurbishment contracts after the contract has been awarded creates problems for clients relating to the overall budget they have to spend. Cost consultants will advise clients who are planning to commission refurbishment work to include a large sum of money to allow for contingencies such as those stated previously. This may mean that the planned scheme may have to be based on an inferior specification in order for the

Figure 1.3
Frequency distribution of
contracts completed above
or below tender price.

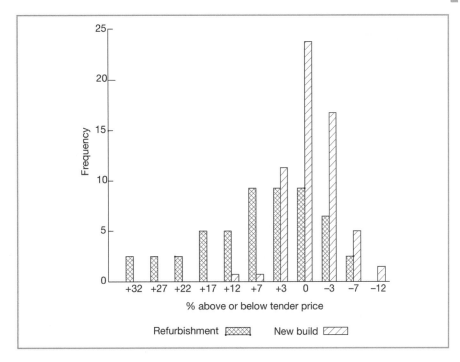

work to be completed within budget. However, none of the contingencies may actually materialise, and the client will have 'spare' money that could have been spent on the scheme. This is not generally the situation in new build work.

Figure 1.3 illustrates the frequency distribution with which build and refurbishment contracts are completed above or below the tender price.

A large percentage of new build contracts are completed for the tender price and a significant number are completed below tender price. Far fewer refurbishment contracts are completed for the tender price, and there are a number that go over tender price significantly. If this occurred and the client had not allowed for contingencies then the work would undoubtedly go uncompleted or the scope of the planned scheme would be significantly reduced. Alternatively, the quality of the scheme would have been reviewed and lower quality products used.

Reflective summary

- A common reason for clients requiring refurbishment of existing buildings is that the building structure is sound but the plan layout of the building is unsuitable for modem purposes.
- There is a common misconception that to refurbish will cost less than to demolish and build new.
- Over half of the value of construction work to buildings undertaken in the UK is used in refurbishment.
- There are a number of factors that can make it difficult to price refurbishment work accurately. Clients are therefore generally advised to set aside a sum of money for any contingencies that may arise during the works.
- There is a far greater tendency for refurbishment contracts to be completed over the tender price than new build contracts. A significant number of new build contracts are completed for the tender price or below.

Review task

Produce a list of the factors that may incur additional costs during refurbishment work.

1.3 | Issues that affect the decision to refurbish

Introduction

- After studying this section you should have developed an understanding of why clients may choose to refurbish a building as opposed to demolishing and rebuilding.
- You should also be able to explain how and why buildings fail, and how these factors are linked to definitions of the size of refurbishment schemes.
- You should also be able to explain the need for appraisal of buildings which may be refurbished, the stages of the appraisal process, and how this can lead to a scheme being deemed as feasible or unfeasible.
- You should also understand the cost implications of undertaking an appraisal of an existing building, and how this may affect the decision of a client to pursue the refurbishment option.

Overview

The decision to refurbish a building is made when a building is not deemed to be 'fit for purpose'. Buildings that may be refurbished are:

- Occupied by the client who wishes to remain in the building, but the building does not suit current business practices
- Existing buildings that are chosen by clients specifically because of location and size, but which are not suitable for the proposed use
- Buildings bought by a developer without a specific occupier in mind. The aim is to refurbish the building and then rent out or sell commercial/industrial/residential space.

Generally, the layouts of older buildings do not suit modern requirements and this is a major reason to refurbish, but refurbishment is also required where buildings have failed.

Failure of buildings

Buildings may fail for a number of reasons. Figure 1.4 shows a breakdown of the most common reasons for this.

Figure 1.4
Causes of failure in
buildings.

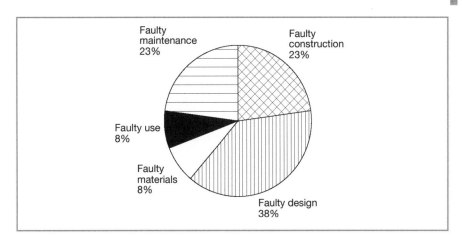

The Building Research
Establishment does produce
prototypes of buildings and
building elements in an
attempt to reduce design-
caused defects.

Faulty design accounts for a significant number of building failures, but this is not a representation of the poor performance of building designers. The nature of construction work does not enable the building of prototypes for testing in every conceivable situation. If we compare the production of cars, for example, a car is designed and then built; it is tested exhaustively, faults identified and then rectified. The model is then tested again and again, and the cycle repeated until all faults are eliminated. If you relate this to building work, you would need to construct a building, test it over time and then demolish and rebuild, removing all the problems in the next design. This is obviously not feasible, and for this reason most designers will specify systems of construction that have been tried and tested over the years. This does, however, lead to a restriction in the amount of 'design flair' that can be incorporated into building design, and sometimes designers will specify the use of a system that has not been tried and tested. These systems can fail either quickly or over a longer period of time, but essentially if these new systems were not to be tried the built environment would be far less interesting aesthetically.

Faulty construction accounts for many building failures and this can be linked to the above. If a new system has not been specified or used previously, then the builder will have no experience of this system and may build it incorrectly. Sometimes lack of suitably qualified supervision can lead to this problem, as can the lack of testing that is carried out during construction work, such as concrete tests. An example of systems that have not been used before creating this problem is that of precast concrete structural frames. During the 1950s, in order to increase the amount of housing, the development of high-rise blocks of flats became endemic. As precast concrete frames were used extensively and successfully in mainland Europe, they were adopted for use as the structural frame in these flats in the UK. However, builders and supervisors had no experience of this type of construction and problems have arisen – or more specifically with the *in situ* concrete joints used to form the joints between precast concrete members. Water has penetrated the joints and corroded the steel reinforcement, which in turn has led to spalling of the concrete around the steel, so that the joints are no longer rigid. Some of these structures have been demolished, but some are still standing and essentially it is only the sheer weight of the structure that holds them together. The term *buildability* is commonly used these days, and basically means that designs have been developed from the perspective of the person who is to construct them and that the details should be very clear. Furthermore, the increasing demand for quality from clients is leading to better qualified supervision of construction work. This should lead to fewer problems of faulty construction and, very importantly from a health and safety perspective, the design should be able to be constructed with a reduction of risk to site workers.

Faulty maintenance accounts for a similar number of building failures, and this can be broken down into two parts: maintenance that has been carried out incorrectly, or more commonly where no maintenance has been carried out during the life of the building. A large section of this book is dedicated to building maintenance, and how it can lead to the improved performance of buildings over their lifespan. If the procedures specified in this book are adopted then this figure should be reduced. However, maintaining buildings costs money, and therefore although building maintenance can be planned and specified correctly, if the funding available is not adequate this will ultimately lead to building failure.

Faulty materials account for fewer, but still substantial, amounts of building failures, and the reasons for this are to a certain extent the same as for faulty design. We cannot test all materials for 60 years before they are used for construction, and we cannot test all materials in conjunction with all of the materials that they may potentially come into contact with. However, as a general rule, materials that are manufactured in factories will be of better quality than materials manufactured on site (*in situ*). More prefabrication should reduce this problem, but it can be argued that prefabrication reduces design flair and flexibility.

Finally, faulty use accounts for some building failures, and this generally occurs where the building is not being used for the purpose for which it was designed. For example, occupiers may wish to create more space and therefore knock down walls without the advice of the designer, which can have major implications and possibly cause a collapse.

Building obsolescence

The physical and functional characteristics of buildings and their condition at a point in time will affect their usability and their desirability to commercial occupiers. In recent times there have been numerous large-scale urban regeneration projects that have incorporated the demolition and replacement of buildings that may be deemed to be in reasonable physical condition. Most major towns and cities display an array of building types, varying in age, form and condition. Many of these are empty or are used for purposes other than that for which they were designed. Whilst these are often in reasonable physical condition and may be capable of refurbishment, they are often considered as obsolete. Key drivers in the decision regarding whether a building is obsolete include physical form and condition, but also increasingly important are aspects associated with the capacity to accommodate modern business technology and the degree to which existing fabric can be upgraded to meet the increasing requirements for carbon reduction and energy conservation.

There are also a number of intangible issues such as the changing trends and fashions for building styles and the very real issue of physical location. A building that is in good condition but that is in the wrong location may be every bit as obsolete as a deteriorated building. It may seem odd that a building might suddenly be in the 'wrong place' but the major urban regeneration projects such as 'Liverpool 1' have impacted upon the layout and zoning of cities and the evolution and relocation of the commercial core of a city have occurred many times.

There have historically been two schools of thought on dealing with building obsolescence at the design and development stage. Firstly there is the view that since buildings will start to deteriorate and become obsolescent shortly after construction, a short design life with the capacity for easy demolition and renewal should be considered. In the 1960s and 70s this was a strongly held view, since the pace of technological advance meant that buildings would become outdated very quickly. At that time there was little drive for sustainable building design and use. The alternative view favoured 'loose fit' building design with the ability to remodel and improve buildings through their lives, thus extending the usable lifespan.

The problem with both of these views is that they deal relatively well with the process of building deterioration but in terms of dealing with obsolescence they cannot cater for the non-physical and non-predictable issues. Obsolescence is different from deterioration in several key aspects. Deterioration is essentially a physical process, whereby the condition of the structure, fabric or services falls below an acceptable level; this is predictable and manageable. However, obsolescence is more amorphous and can result from a range of non-physical, non-predictable factors. The Royal Institution of Chartered Surveyors (RICS) identifies the factors that may lead to obsolescence to include the following:

- *Technological factors*: the increasing use of technology for business and the ability of a building to accommodate the supporting infrastructure.
- *Functional/social factors*: the changing nature of society, business and building use means that some building types are simply no longer needed in the volume that they once were. The large numbers of churches that have been converted to other uses well illustrates this.
- *Economic factors*: buildings are financial assets, once their income generation potential reduces, or their function becomes possible in a cheaper way, they will cease to be viable assets.
- *Physical/legal factors*: the ability for older buildings to meet the requirements for energy conservation, accessibility and the control/removal of deleterious materials may simply be uneconomical or undeliverable.
- *Aesthetic factors*: some buildings simply appear to be too old, too ugly or too out of context to be viable refurbishment propositions.

The decision to refurbish as opposed to demolish and rebuild

The decision to refurbish a building as opposed to demolishing any existing building and undertaking a new build scheme may be taken for the following reasons:

- Refurbishment is currently seen as a more sustainable option than new build work (see Section 1.4)
- Buildings may be listed (see Section 1.5)
- Many buildings are structurally sound, and therefore to demolish a sound structure is not economically viable
- The current or proposed occupier may wish to change the use of the existing building
- The existing building services may not provide the levels of performance that are required by building occupiers, and work may be required to the structure in order to facilitate new services.

Refurbishment schemes can be classed as:

- *Minor refurbishment*: this would include replacement or upgrading of plant and services, redecoration and new floor coverings.
- *Major refurbishment*: this would include replacement of major plant and services, suspended ceilings, floor finishes, raised floors and internal walls.
- *Complete refurbishment*: only the substructure, superstructure and floor structure are retained.
- *Redevelopment*: where the only element to remain is the existing façade and foundations to the façade.

If planned maintenance is carried out on buildings, the need for large-scale refurbishment should be reduced.

The extent of refurbishment required would be established after an appraisal of the existing building has been carried out. In the initial appraisal the following assessments would be made:

- Whether the building is in a state of serious deterioration, and a collapse is possible
- Whether the building is suffering from significant deterioration, which may indicate that major remedial works are necessary, i.e. works to the structure
- Whether or not there are any evident defects in the original design and/or construction that have caused or are causing damage
- Whether or not there has been any accidental damage to the building
- Whether it is feasible that the building could be used for an intended change of use
- Whether a further and more detailed structural survey is required.

The process to be adopted when undertaking an appraisal is shown in Figure 1.5. As the figure shows, there is a great deal of work that is required to be undertaken before any drawings are prepared. This has a major cost implication that does not occur when specifying new build work. In new build work the only real investi-

Figure 1.5
The appraisal process.

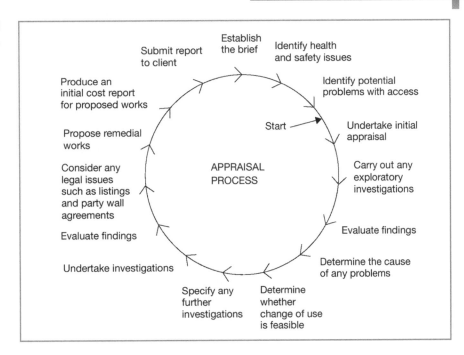

gation that is required is a soil investigation to allow for the foundations to be designed. The additional cost of the appraisal needs to be borne by the client, who may be deterred from choosing to refurbish because of this cost. The appraisal could be completed and the report could recommend that refurbishment is not a feasible option and therefore the money spent is unrecoverable. However, there may be grants available to refurbish buildings that are not available for new build schemes and this may provide an incentive to clients to pursue the refurbishment route.

Reflective summary

- The decision to refurbish a building is made when a building is not deemed to be 'fit for purpose'.
- Generally, the layouts of older buildings do not suit modern requirements. This is a major reason to refurbish, but refurbishment is also required where buildings have failed.
- Buldings may fail because of faulty initial design, construction, use, maintenance or materials. Commonly a combination of these leads to building failure.
- Returbishment work can be classified by the extent of work required to make it 'fit for purpose'.
- An appraisal of an existing building is essential when deciding whether a refurbishment scheme is feasible and viable.
- The cost of undertaking an appraisal may deter clients from pursuing the potential of a refurbishment scheme.

Review task

Produce a list of how buildings can fail and give examples in each case.

Outline the procedures that need to be undertaken when carrying out an appraisal of an existing building.

1.4 | Refurbishment vs. redevelopment from an environmental perspective

Introduction

■ After studying this section you should have developed a basic understanding of the environmental impact of construction work, both new build and refurbishment. However if you wish to gain a greater understanding of this, chapter 14 in *Construction Technology 2* goes into much greater detail.

■ You should also understand the difference between the environmental impact of refurbishment work compared to new build work and be able to illustrate the differences in adopting principles of sustainable design between the two different types of construction project.

■ You should also have developed a basic understanding of the implications of complying with Part L of the Building Regulations (England and Wales) and Part J of the Building Technical Standards (Scotland) in relation to refurbishment work and possible implications for future buildings.

■ You will also have developed a knowledge of how buildings can be assessed as to their environmental impact both during construction and during the life of the building.

Overview

Chapter 14 in *Construction Technology 2* outlines the need for construction work and buildings to be more sustainable in order to protect the planet from further global warming and excessive pollution. The main principles of 'green buildings' are also explained but in the context of new build work as opposed to refurbishment work. Buildings are believed to account for 50 per cent of all energy used globally. Half of that energy is consumed during the construction of the building and half during the life of the building (Figure 1.6). It is obviously therefore important that materials and systems of construction that can reduce this energy use are utilised by the construction industry as much as possible. Refurbishment, rather than redevelopment, is currently seen to be the more sustainable option because the amount of new build work is reduced. However there are drawbacks:

■ The green building solutions available for refurbishment work are more limited than those for new build work

■ It is more difficult to comply with Part L of the Building Regulations for refurbishment work than for new build

■ The shape and form of many existing buildings are not conducive to conversion to modern-day occupant requirements that promote building efficiency.

The UK government has pledged to contain and reduce the growth in carbon dioxide emissions, a large part of which is due to the energy use in buildings. The

Figure 1.6
Energy used by the built
environment.

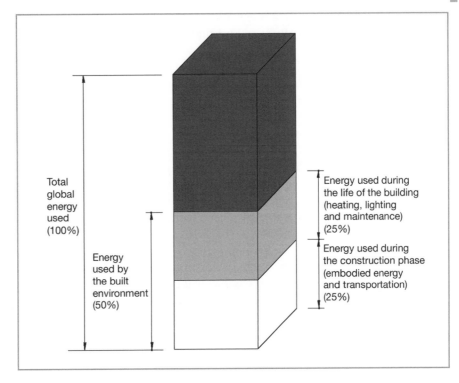

energy consumption of houses represents 27 per cent of the national emission of carbon dioxide, and a further 19 per cent comes from non-domestic buildings. In refurbishment projects the services installation is generally a retrofit system which is a system designed to fit into an existing building rather than a new build system where the design of the building and the services installation are viewed in a more holistic way. The options therefore available in new build work are greater and more efficient than those available for refurbishment projects. The principles of upgrading and retrofitting of building services are detailed in Chapter 4 (Section 4.6) and the principles are the same if a more sustainable building is required. Some of the solutions may be different though.

Reducing environmental damage during the construction stage involves utilising the principles of sustainable construction. These principles can be summarised as:

An EPM is basically a system that allows for the comparison of alternative materials from an environmental perspective.

- Choosing materials and systems using an environmental preference method (EPM).
- Non-sustainable materials and systems must only be used when there is no feasible alternative
- Materials must be reused wherever possible
- The amount of waste water that is generated on-site must be reduced so as to prevent pollution of the surrounding areas
- Excavation methods must be carefully considered
- Waste must be sorted

- Storm water must be contained
- Waste and tipping must be minimised.

Utilising these principles it becomes apparent why refurbishment is the more sustainable option. If less new material is used during the construction phase then fewer materials need to be assessed using an environmental preference method. Excavation in most refurbishment projects is not required as the existing substructure will remain, and waste and tipping are reduced because demolition is limited. There is also the possibility of reusing any demolished materials from the site to be refurbished, but it is more likely that reclaimed materials can be used from other demolished buildings to try to 'match' older materials on the refurbishment project. An example of this is replacing roof tiles on older buildings. Tiles that have been reclaimed from demolished buildings can be used and newly manufactured materials will not need to be specified.

The Building Research Establishment (BRE) has developed a tool that enables users to compare refurbishment and redevelopment scenarios for a particular site from an environmental perspective. This tool is called Office Scorer and allows for the comparison of environmental impacts and whole life costs of refurbishment and redevelopment. The BRE recommends the early use of this tool in the decision-making process, when the options of refurbishing, redeveloping on the same site and redeveloping elsewhere are being considered. It uses financial costs (£) or ecopoints per m^2 or per person as the basis for comparison. One hundred ecopoints are equal to the environmental impact of one person in the UK over one year.

A major part of this tool looks to identify the amount of embodied energy required in material and systems manufacture. The amount of embodied energy of a material is determined after assessing the environmental impact of the manufacture of materials and transportation, the construction process (including transport to site, any maintenance repairs or replacement required over the life of the

Figure 1.7
Comparison of the environmental impact of refurbishment vs. new build.

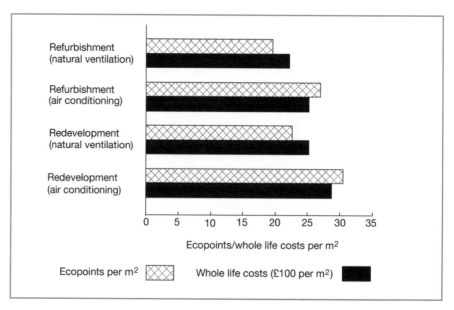

building), and the energy required for demolition and disposal at the end of the life of the building. The research undertaken by the BRE suggests that from an environmental perspective, refurbishment is the best option. Figure 1.7 compares the ecopoints and costs per m^2 of undertaking refurbishment and redevelopment. It also shows that if the scheme is to include air conditioning the ecopoint score will be higher than if natural ventilation is used. This indicates that refurbishment is indeed the better option, and if natural ventilation is adopted this is even better from an environmental perspective.

Design and construction issues related to the changes to Part L of the regulations

When considering refurbishment of a building, there are a number of issues that need consideration.

One of the key issues for modern buildings is the ability to minimise heat loss through air leakage from the enclosure. Current regulations have increased the significance of this factor and refurbished buildings are sometimes less well able to cope with the requirements for minimisation of air leakage than are new buildings.

Airtightness could become the 'Achilles' heel', as designers will be required to ensure that a building meets the air leakage targets. Mobile testing units can be used by building control bodies and if buildings 'fail' they will have to be adapted to ensure compliance. Robust standard details will have to be produced, but they may require improved standards of workmanship. Current industry operatives may need retraining. Building control bodies will have to monitor compliance and may need extra resources. This may create problems in refurbishment schemes and may deter designers from choosing the refurbishment option because of a fear that the building may not comply with this regulation when complete. Further major works may be required which will require additional funding.

The revisions discourage reliance on mechanical cooling systems and encourage the reduction of overheating through shading, orientation, thermal mass, night cooling etc. There can be a trade-off between the elemental U values and the efficiency of the boilers, but this could lead to constantly changing to boilers that are increasingly more efficient, and could lead to reliance on boilers to achieve compliance. In addition, every building will need some low-energy lighting systems.

Refurbishment projects may not be able to comply with the regulations and design and build as a procurement tool may drop in popularity due to the amount of work required during the pre-tender design stage by the contractor, which could prove financially uneconomical. However, the changes are excellent for the sustainability-minded designer and design and build contractors.

Achieving sustainable design in refurbishment work

The principles of sustainable design have been outlined in *Construction Technology 2* and the social, economic and ecological benefits of adopting these principles are

Table 1.1 Potential benefits/pitfalls of refurbishment vs. new build work to achieve sustainable design.

	Potential benefit	Potential pitfall
Water	Natural water courses should be preserved as the building is existing and there will be no disturbance to the ecosystem	It may not be possible to reduce the amount of potable water used due to restrictions within the existing structure
Energy	If the existing structure can be better insulated then energy use can be reduced	It may not be possible to install more energy-efficient heating and cooling systems due to restrictions within the existing building structure. It may not be possible to reconfigure the existing structure to allow for more natural light or stack cooling systems etc.
Materials	The mere fact that the building is being refurbished may positively encourage the reuse of materials to 'match' the existing building. If the building is old then the materials may be natural and locally sourced, which is a major principle of sustainable design	In order to achieve the levels of insulation and U values required by Part L of the Building Regulations, specialist materials may be required that are not locally sourced. This may require significant embodiment and produce pollution in manufacture
Siting	It may be possible to resite the building orientation dependent on the extent of the refurbishment	Dependent on the extent of refurbishment the siting may be impossible to change. Planning restrictions and conservation laws may reduce the potential to reorientate the building
Operation	Installed systems will definitely be designed to use less energy and reduce air pollution and therefore any refurbishment will be of benefit	The potential choice of systems may be reduced and the ability to design a services installation alongside the structure with basically a clean sheet is reduced. The chances of designing an intelligent building are seriously limited

U values are measured in W/m^2 K. A U value of 0.25 means that one quarter of a watt of heat is lost through each square metre of the element when a 1°C temperature difference exists between the inside and outside of the element.

discussed. Table 1.1 illustrates the potential benefits and pitfalls of refurbishment as opposed to new build work in achieving a sustainable design.

Of note before referring to the table is that the extent of refurbishment will also impact on these principles. A refurbishment project could be something as basic as the replacement of original windows to the complete demolition of the entire building except for one façade. As a rule of thumb the more extensive the refurbishment, the greater the ability to comply fully with principles of sustainable design.

Facilities management issues

All buildings will have to have energy meters and building log books, which are essential if building owners and operators are to monitor energy consumption against a benchmark figure.

The changes to the regulations have been made in order to reduce the negative environmental impact of buildings. However, there is a danger that the regulations will lead to a reduction in refurbishment projects because of the concern over non-compliance with the regulations when work is complete, and the high levels of funding required to ensure compliance. From a designer's perspective it will be easier to guarantee compliance for new build schemes than for refurbishment schemes. If this occurs then the regulations may actually work against environmental improvements, as it is clear that from a sustainable angle, refurbishment is the better option as less new material is required and therefore the amount of embodied energy used will be reduced.

Carbon reduction and energy performance

In the UK there are several initiatives aimed at improving the environmental performance of existing homes through 'retrofit' projects. Nearly 90 per cent of the homes that will exist in 2050 have already been built.

As building users become more aware of the need to conserve energy and the implications of higher energy consumption in terms of cost they will become more discerning in their selection of commercial property. In the UK, Energy Performance Certificates (EPCs) have been introduced to promote the improvement of energy performance of buildings and to provide end users with informed choice when selecting commercial property. The introduction of EPCs is a response in England and Wales to the European Directive on the Energy Performance of Buildings.

EPCs provide information regarding a commercial building's energy performance and provide an energy rating for the building that is akin to those that have been established for several years in the domestic appliance sector. They aim to provide information to potential buyers or tenants about energy performance of a building, so that they can make informed judgements about buying or occupying. The energy rating set out within the EPC is based on the performance potential of the building fabric and services. The rating given is based on the energy performance of the building relative to a benchmark which can be used to make comparative assessments of different properties. The EPC report also makes recommendations on how the energy performance of the building could be improved, supported by broad information regarding possible payback periods.

The certificate provides a rating from A to G, where A is most efficient and G is the least efficient. The better the rating, the more energy-efficient the building is, and the lower the fuel bills are likely to be. The energy performance of the building is shown as a carbon dioxide (CO_2) based index, reflecting the increasing focus upon 'carbon reduction' rather than simply 'energy conservation'. For public buildings and facilities, Display Energy Certificates (DECs) are required to be displayed in areas visible to public users to indicate the energy performance of these buildings.

Carbon Reduction Commitment (CRC)

The greatly increased awareness of sustainability issues in the construction and operation of commercial property has been supported by legislation and government-driven initiatives that force building owners and users to consider energy performance in use. One of the manifestations of this in the UK has been the creation of the Carbon Reduction Commitment programme. This forces large consumers of energy in commercial buildings to actively seek to reduce the carbon emissions arising from energy use on an ongoing basis.

The scheme requires annual monitoring of energy use and is essentially a mandatory emissions trading scheme that relates to energy consumers above a defined consumption level. The scheme allows organisations to buy allowances at a cost per tonne of CO_2 to cover the equivalent amount of CO_2 produced by the organisation within the year.

The criterion for inclusion within the Scheme is that in excess of 6000 MWh of electricity was consumed through half-hourly meters in 2008. Organisations will be required to forecast and report CO_2 emissions annually and to purchase allowances accordingly. Although metered electricity is the defining element for inclusion in the scheme, all energy sources associated with fixed facilities will need to be accounted for through purchase of allowances. Allowances will be surrendered at the end of each year to cover reported and evidenced CO_2 outputs.

The scheme will operate on the basis of measurables or 'metrics' that allow organisations to measure and report emissions and, ideally, improvements in energy conservation.

Three metrics are defined within the Scheme as follows:

- *Absolute Metric:*
  This measures and reports the change in yearly emissions relative to an average of the prior 5-yearly period.

- *Early Action Metric:*
  This is a measure of action taken prior to the start of the scheme and is based on the percentage of energy used covered by voluntary automatic meter readings and the percentage of emissions covered by the Carbon Trust Standard.

- *Growth Metric:*
  This relates to change in emissions per unit of organisations' turnover.

Organisations will be required to report on their annual energy consumption and there are significant penalties for non-compliance. All CRC participants will be required to self-certify their data and reports will need to be submitted on-line to the scheme administrator.

The actions that are required for participation and reporting are set out below:

Stage 1: Establish and record CRC 'responsibility chain'

It is necessary to establish and define a responsibility chain for CRC within the organisation. This should comprise the following:

- *Organisation Level*
  It is a requirement of the scheme to identify a CRC Coordinator within the highest level of the parent company. Ideally this should be a Board-level position, and have responsibility and accountability for participation in the scheme.

- *Building Level*
  It is a requirement to identify a CRC officer for each building. The CRC officer should act as a contact point for the building and should be responsible for identifying and progressing energy improvements in the building.

Stage 2: Monitoring and recording

Between 1st April and 31st March each year it is required to monitor and record all energy sources and usage. The first year's data will then be used to prepare a footprint report.

Stage 3: Purchase of allowances

The data generated and submitted from the foregoing stages is used to inform the process of buying allowances for consumption for the first year and predicted consumption for the following year. Auctions will then take place annually.

Stage 4: Monitoring and reporting annually

On an ongoing basis, emissions will need to be recorded and reported annually, based on the April to March 'Scheme year'.

Recording and reporting

The energy consumption that needs to be considered and reported should be converted to CO_2 emissions. All energy consumed will be assessed and converted/reported including electricity, gas, fuel oil and other sources.

Data reporting should include the following details:

- Energy supplier
- Unique meter ID/Code
- Reading type (Actual, Automated, Self-read, Estimated)
- Total annual energy usage through meter.

The following exclusions apply to the scheme:

- Domestic energy use
- Energy associated with transport
- Any unconsumed supplies.

Evidence pack

The Scheme requires the use of an evidence pack that incorporates information defined by the scheme administrator. The following information is captured to allow the generation of a 'footprint report' and subsequent annual reporting:

1: *Structural records*
   This section defines the scope of operation of the organisation, the types of site occupied and the types of energy used.

2: *Data records*
   This section allows recording of energy supplies by type and by supplier and should be supported by inclusion of copies of invoices and statements together with meter readings for the various energy sources and types.

3: *Special event records*
   This section allows the recording of special events such as energy disruptions, changes of suppliers, leakages etc.

4: *Data on early actions taken*
   This section allows early actions associated with the scheme to be recorded.

5: *Records relating to exemptions*
   This section allows data to be recorded/reported that may be used in support of claims for exemptions and energy credits.

Reflective summary

- Buildings are believed to account for 50 per cent of all energy used globally. Half of that energy is consumed during the construction of the building and half during the life of the building.
- Refurbishment, rather than redevelopment, is currently seen to be the more sustainable option because the amount of new build work is reduced.
- Part L of the Building Regulations (England and Wales) and Part J of the Building Technical Standards (Scotland) deal with the energy efficiency of buildings.
- There are a number of methods to prove compliance with the regulations, both for domestic and for industrial and commercial buildings.
- The changes to the regulations have been made in order to reduce the negative environmental impact of buildings. However, there is a danger that the regulations will lead to a reduction in refurbishment projects because of the concern over non-compliance with the regulations when work is complete.
- The level to which sustainable design principles can be adopted in refurbishment work generally Increases as the extent of the refurbishment scheme becomes greater.
- The increased focus on carbon reduction is supported by government initiatives such as the Carbon Reduction Scheme and the requirement for EPCs and DECs.

Review task

Outline the methods for testing that buildings comply with Part L of the Building Regulations.

Produce a matrix to compare the environmental impact of refurbishment schemes compared to new build schemes. This should be developed by identifying criteria to compare the two types of construction work and then grading the criteria.

With regard to EPCs and DECs, produce a list of the factors that might affect the intrinsic energy performance of a building from the points of view of fabric and services.

1.5 Overview of statutory control of buildings

Introduction

- After studying this section you should be able to appreciate the evolutionary development of building control through public health and other associated legislation.
- You should be able to outline the basic processes required to obtain formal approval to build and how control extends to the building process.
- In addition you should appreciate the implications of undertaking refurbishment works in the context of adjoining landowners and the legislative tools that exist to control this.
- You should appreciate the relevance of listed building status in the context of building refurbishment.
- You should also appreciate the basis of statutory control of listed buildings and the mechanisms by which this can be imposed by various agencies.
- In addition, you should understand the basis upon which listed building status is granted and the broad classifications used to describe different grades of listing.
- Building to accommodate disabled people is now a major issue, and an outline of the requirements is given in an overview of Part M of the Building Regulations.
- Standards are considered, and you should be able to develop an understanding of the future changes to the regulations that may occur and the implications for future buildings.
- You should also develop an understanding of the legal responsibilities of all the parties involved in the construction process with regard to the health, safety and welfare of site workers and the public during construction work, and develop an understanding of relevant regulations.

Overview

The application of controls over the building process is a relatively new concept. Prior to the first set of Building Regulations in 1965, control over the building process was limited. The first real major realisation of the need for control probably came in the aftermath of the Great Fire of London in 1666, and in 1667 the building of the external structure and fabric of buildings in timber was outlawed in London.

A main feature of the development of control in the rest of England was the various Public Health Acts (PHA), and probably the most notable of these was the PHA 1875. This had three main focuses: structural stability, dampness and sanitation. Following this Act a set of Model Bye-Laws was issued in 1877 as a guide for Local Authorities who were delegated the responsibility for setting and enforcing minimum standards of construction.

Following limited progress in the 19th century, legislation concerning construction was limited, and mainly in the area of public health. By the 1950s many local authorities were issuing bye-laws peculiar to their locality, and this made it difficult because laws varied from area to area. Because of the need for consistency, National Bye-Laws were established in 1952 and, following these, items of significance include the Public Health Act 1961 and the first set of national Building Regulations in 1965. The aims of the Building Regulations were to set out what the minimum standards acceptable for building works are. It is not within the scope of this book to discuss all the Building Regulations, but reference is made to them throughout the text. An overview of Part M is given in this section.

A major reference for the safety of persons associated with the building process is the Health and Safety at Work Act 1974, which applies to all industries and aspects of construction work. The Act is an 'umbrella' for a huge set of regulations that are specific to individual industries and aspects of those industries. Some of the most recent and important regulations affecting construction work will be discussed in this section.

Listed buildings are those included in the statutory lists of buildings of special architectural or historic interest. Detailed records of listings are held which define the nature and extent of the listing and hence the degree to which the listing affects the potential to alter, refurbish or demolish the building.

Many of the older buildings in the United Kingdom, as well as some newer ones, are listed. The listed status of a building indicates that it is thought to be worthy of special protection for the future. The 'List of Buildings of Special Architectural or Historic Interest' is compiled by national conservation agencies and the Department for Culture, Media and Sport. Although it is possible to list a building which was constructed as recently as 30 years ago, in practice the majority are much older.

A building can be listed at any time, and it is often the case that building owners are notified of this process after the event by the Department of the Environment and the Local Authority. Under normal circumstances there is no right of appeal against listing. Naturally this can have major implications for any proposals to refurbish, alter or demolish existing buildings and the formal process of listed building consent must be initiated prior to undertaking works. The nature of works allowed on listed buildings is tightly controlled in terms of scope and quality, and the consent to undertake works will often carry with it strict conditions. This has significant cost implications when working with listed buildings.

The Building Technical Standards are the Scottish equivalent of the Building Regulations in England and Wales.

Implications of listed building status

As previously noted, the controls that are applied to listed buildings are far stricter than normal planning controls that apply to other buildings. This does not mean that changes cannot be made to listed buildings, but it does imply a requirement to apply for Listed Building Consent to carry out any works that would affect the character and/or architectural merit.

If a listed building is demolished, altered or extended without consent, such that the character and appearance of the building are affected, the Local Authority may serve a Listed Building Enforcement Notice requiring reinstatement of the building as it was prior to the work. If this is not possible, the building owner may be fined, imprisoned, or both. In practice, most repair, maintenance and alteration to listed buildings requires Listed Building Consent. Work which involves replacing building elements, components and finishes in different styles or materials from the original will need consent, as will demolition works and extension or alteration. Whilst the listed elements of many buildings are restricted to the external elevations (allowing façade retention schemes to be initiated), it is important to remember that the interiors of some buildings are also protected. Architectural features of the building interior, such as staircases, may often need to be retained. Boundaries to properties may also form part of the listed building, and work to these elements may also require consent.

The vast majority of listed buildings are of considerable age and the need for ongoing repair and maintenance increases as they get older. In some situations the building may fall into a state of disrepair such that its special interest as a listed building is in danger. In such situations the Local Authority may be forced to serve a 'Repairs Notice'. The notice will specify work that should be undertaken to bring the building up to an acceptable condition. If the required work is not carried out within a defined period, the Local Authority may acquire the building by compulsory purchase. It also has the power to undertake works and claim the cost from the owner.

The nature of listed buildings

Listed buildings in England and Wales are classified by grade to define their relative importance with reference to historic or architectural merit. The grades currently adopted are as follows:

- Grade I: buildings of exceptional interest
- Grade II*: particularly important buildings of more than special interest
- Grade II: buildings of special interest, which warrant every effort being made to preserve them (these account for the vast majority of listed buildings).

The process of listing applies a range of broad rules to assist selection. Although these are by no means definitive, they include the following general principles. Buildings likely to be listed include:

- Most buildings built before 1700
- Selected buildings built between 1700 and 1840
- Buildings of definite quality and character built between 1840 and 1914
- Selected buildings of high quality built between 1914 and 1939
- Outstanding buildings of exceptional interest built after 1939.

Buildings become eligible for listing when they are 30 years old; however, in exceptional circumstances buildings 10 years old which are deemed to be under threat may be listed.

In Scotland, listed buildings are dealt with by grading using a different system including categories A, B and C(S):

- Category A includes buildings of national architectural or historic importance
- Category B includes buildings of local importance
- Category C(S) includes buildings which may have been altered, or in some cases buildings of little individual merit but which group well with others in Categories A or B.

In Northern Ireland buildings of historical or architectural merit may be listed but this is non-statutory.

Modern, well-kept buildings are easy and cheap to keep and manage. However, older buildings become increasingly difficult to maintain as they age. The physical deterioration of the structure and fabric will become more extensive with time, and once decay is allowed to set in a building becomes increasingly difficult and expensive to maintain. When considering refurbishment and alteration, the age of the basic building must be considered. If the building is to be fit for alternative use, refurbishment will be required, and if the disposal value is to be maintained it must be properly maintained. When dealing with older buildings it is essential that a long-term view of the cost-effectiveness of maintenance and refurbishment be taken.

Figure 1.8 shows examples of listed buildings, one old and one more modern. They are both listed for specific reasons.

Figure 1.8
Two listed buildings side by side.

Refurbishing buildings for people with disabilities

Physical disabilities include mobility problems, impaired hearing and sight.

With an increasingly ageing population, the proportion with physical disabilities is a factor of increasing significance. The need to improve access to buildings is reflected in the development of regulations concerning providing suitable facilities for people with disabilities.

The principal legislation which concerns provision for the physically disabled is contained in the Disability Discrimination Act 1995 and Approved Document M of the Building Regulations.

The checklists given later in this section are examples that offer guidance on adapting housing and refurbishing offices to accommodate the physically disabled. Their application will hopefully assist in evaluating the suitability of existing buildings to cater for the special needs of physically disabled people. They will also assist in identifying the building elements which may need changing to provide suitable access.

Before any design is undertaken, the architect could use these guidelines to assess the existing building, identify where there is non-compliance and design in solutions to these problems. It may be deemed that it is impossible to comply with the regulations when undertaking a minor refurbishment and a more major scheme may be required. This will have a cost implication for the client, who may wish to choose another building to refurbish. Occasionally, it may be impossible to convert an existing building to comply with the regulations, and demolition and new build may be the only option. The checklists can also be used to check that any new build scheme is designed in accordance with statutory requirements.

The statutory requirements concerning the provision of suitable adaptations of refurbishment of buildings are contained in the Building Regulations 2000, Access and Facilities for Disabled People – approved document M. Part M offers guidance on where the requirements apply, new buildings, extensions, alterations and external features. The specific matters dealt with are summarised as follows:

- Section 1: Means of access to and into buildings other than dwellings
- Section 2: Means of access within buildings other than dwellings
- Section 3: Use of buildings other than dwellings
- Section 4: Sanitary conveniences in buildings other than dwellings
- Section 5: Audience or spectator seating in buildings other than dwellings
- Section 6: Means of access to and into the dwelling
- Section 7: Circulation within the entrance storey of the building
- Section 8: Accessible switches and socket outlets in the dwelling
- Section 9: Passenger lifts and common stairs in blocks of flats
- Section 10: WC provision in the entrance storey of the building.

Examples of checklists

The aim of using these checklists is to try to achieve as many positive responses as possible.

These checklists can be used for assessing existing buildings, and for informing the design of new build work.

ENTRANCE	Yes	No
Is there at least one primary entrance usable without assistance for someone in a wheelchair?	☐	☐
Is the main entrance accessible from the outside without having to use steps or stairs?	☐	☐
Does the entrance have a clear opening of at least 810 mm?	☐	☐
Does the entrance have a vision panel 900–1500 mm from the finished floor level?	☐	☐
Is there a ramp outside the main entrance?	☐	☐
Are there handrails at the ramp at a height of 800–920 mm?	☐	☐
Is there a level, paved area at least 1500 × 1500 mm outside the door?	☐	☐
Is the threshold less than 15 mm high?	☐	☐
Is the accessible entrance identified by the international symbol of access?	☐	☐
Is there an automatic door-opening device?	☐	☐
Does the door have a clear opening of at least 810 mm?	☐	☐
Are the door handles 760–915 mm from the floor?	☐	☐
Is there space to manoeuvre a wheelchair in the vestibule?	☐	☐
Are the floor surfaces slip-resistant?	☐	☐
Is the doorbell or call button lower than 900 mm?	☐	☐
Where revolving doors are used, is there a clearly marked alternative route not less than 900 mm wide?	☐	☐
Are doors operable by a single effort?	☐	☐
Do the door closers allow for the use of the doors by disabled persons (delayed action)?	☐	☐
Are the lock and opening mechanisms operable with one hand?	☐	☐

RECEPTION/FOYER	Yes	No
Does the door have a clear opening of at least 810 mm?	☐	☐
Are light switches within the range of 835–1065 mm above the floor?	☐	☐
Is a 300 mm minimum unobstructed space provided next to the leading edge of the doors?	☐	☐
Is there adequate space to manoeuvre a wheelchair in the room?	☐	☐
Is the intercom within a range of 835–1065 mm above the floor?	☐	☐
Is the window cill 760 mm or less from the floor?	☐	☐
Are the window operating devices operable from a wheelchair?	☐	☐
Are the heating/air-conditioning controls operable by a disabled person?	☐	☐

Is the counter height within a range of 835–915 mm? ☐ ☐

Are the floors slip-resistant? ☐ ☐

Does the carpeting allow free movement of wheelchairs? ☐ ☐

Are all surfaces free from glare? ☐ ☐

Is the lighting adequate for a visually impaired person? ☐ ☐

Is the signage adequate for a visually impaired person (size, colour, contrast)? ☐ ☐

Is there proper demarcation of differences in level by contrasting colour and appropriate lighting? ☐ ☐

Is there an induction loop and/or other audio equipment installed for hearing-impaired persons? ☐ ☐

OFFICE/WORKROOMS	Yes	No
Does the door have a clear opening of at least 810 mm?	☐	☐
Are the door handles within a range of 750–915 mm from the floor?	☐	☐
Are light switches within the range of 835–1065 mm above the floor?	☐	☐
Is the thermostat at a height not greater than 1065 mm from the floor?	☐	☐
Is the intercom within a range of 835–1065 mm above the floor?	☐	☐
Is the window cill 760 mm or less from the floor?	☐	☐
Are the window operating devices operable from a wheelchair?	☐	☐
Are other controls operable by a disabled person?	☐	☐
Is the counter height within a range of 835–915 mm?	☐	☐
Is there shelving that is reachable from a seated position?	☐	☐
Is there adequate space to manoeuvre a wheelchair in the room?	☐	☐
Are the floors slip-resistant?	☐	☐
Does the carpeting allow free movement of wheelchairs?	☐	☐
Are all surfaces free from glare?	☐	☐
Is the lighting adequate for a visually impaired person?	☐	☐
Is there proper demarcation of differences in level by contrasting colour and appropriate lighting?	☐	☐
Does a blind person have an unobstructed path without protrusions from walls, floors or elsewhere?	☐	☐
Is there an induction loop and/or other audio equipment installed for hearing-impaired persons?	☐	☐
Is the level of mechanical and other background noises low enough to avoid interference with sound reception on a conversational level by persons using hearing aids (less than 85 dB)?	☐	☐
Is the room accessible by wheelchair from the main entrance?	☐	☐

TOILETS/WASHROOMS	Yes	No
Is the designated toilet/washroom generally accessible from other areas?	☐	☐
Is the designated toilet/washroom appropriately signposted?	☐	☐
Does the door have a clear opening of at least 810 mm?	☐	☐
Are the door handles within a range of 750–915 mm from the floor?	☐	☐
Is there a turning circle 1500 mm in diameter in the room?	☐	☐
Are light switches within the range of 835–1065 mm above the floor?	☐	☐
Are the mirror and shelf less than 960 mm above the floor?	☐	☐
Are towel racks, soap holders and other fixtures in an accessible space less than 1000 mm above the floor and less than 550 mm from the front of the counter?	☐	☐
Is the top of the toilet seat 475 mm above the floor?	☐	☐
Are there adequate plastic-coated grab bars at the toilet?	☐	☐
Is there a space at least 600 mm wide beside the toilet to allow for a lateral transfer?	☐	☐
Is there space for a wheelchair at the front of the toilet (minimum depth 1350 mm)?	☐	☐
If the toilet is in a cubicle, is it at least 1500 mm wide and 1500 mm deep?	☐	☐
Are taps reachable and easily operable from a wheelchair?	☐	☐
Is the toilet flushing device reachable and easily operable from a wheelchair?	☐	☐
Does the room have a device to signal for assistance?	☐	☐
Does the room have space for an attendant assisting someone in a wheelchair?	☐	☐

HORIZONTAL CIRCULATION	Yes	No
Is directional signposting provided in the foyer/reception?	☐	☐
Is directional signposting accessible to the blind?	☐	☐
Are cloakrooms visible from the foyer/reception?	☐	☐
Is the cloakroom counter less than 835 mm high?	☐	☐
Is there an accessible cloakroom with hooks less than 1420 mm high?	☐	☐
Is the corridor at least 1200 mm wide where wheelchairs must pass one another?	☐	☐
Do corridors have slip-resistant floors?	☐	☐
Does the carpeting allow for free movement of the wheelchairs?	☐	☐
Are surfaces clear from glare?	☐	☐
Is there clear demarcation of differences in floor level?	☐	☐

STAIRS	Yes	No
Is there an alternative to stairs?	☐	☐
Are there handrails at a height of 920 mm?	☐	☐
Is there a suitable continuous handrail on each side?	☐	☐
Does the handrail extend at least 300 mm beyond the bottom step?	☐	☐
Are flights at least 1000 mm wide?	☐	☐
Is the rise of flight of stairs between landings not more than 1800 mm wide?	☐	☐
Do the handrails have tactile cues at changes of floor level?	☐	☐
Are the risers less than 170 mm high?	☐	☐
Are treads more than 250 mm high?	☐	☐
Do the treads have a slip-resistant finish or non-slip nosings?	☐	☐
Are the edges clearly marked for visually impaired people?	☐	☐
Have open risers been avoided?	☐	☐
Are the stairs well lit?	☐	☐

RAMPS	Yes	No
Are ramps provided where necessary?	☐	☐
Is the length between landings less than 9000 mm?	☐	☐
Is the ramp width at least 1200 mm?	☐	☐
Is the gradient 1:12 or less?	☐	☐
Is the ramp well lit?	☐	☐

Health, safety and welfare on construction sites

The Health and Safety at Work Act 1974

This is an umbrella for a multitude of regulations regarding the health and safety of employees in the workplace, and covering every industry.

The purpose of the Health and Safety at Work Act is to provide the legislative framework to promote, stimulate and encourage high standards of health and safety at work. The Act is implemented by the Health and Safety Executive and the Health and Safety Commission:

- *Health and Safety Commission*: responsible for analysing data and publishing results regarding accidents, and for proposing new regulations to help solve problems.
- *Health and Safety Executive*: responsible for ensuring that regulations are complied with.

The latter is the body that can issue improvement and prohibition notices, prosecute, and seize, render harmless or destroy any substance which is considered dangerous.

The construction industry has a poor record with regard to health and safety, and many regulations have been introduced which deal specifically with the industry, such as lifting appliances and excavations. High-risk trades include steel erection, demolition, painting, scaffolding, excavations, falsework, maintenance, roofwork and site transport.

The Health and Safety Executive produces checklists for contractors to ensure that all precautions that can be taken to prevent accidents are implemented. Some areas of construction work have to be checked weekly and the findings entered in a register. These are scaffolding, excavations, lifting appliances and cranes. Although there is a large body of law covering many aspects of health and safety at work, we could define the main statutes that are particularly important in construction as:

- The Health and Safety at Work Act 1974
- The Management of Health and Safety at Work Regulations 1999
- The Construction (Design and Management) Regulations 2007
- The Work at Height Regulations 2005
- The Manual Handling Operations Regulations 1992
- The Provision and Use of Work Equipment Regulations 1998 (PUWER)
- The Lifting Operations and Lifting Equipment Regulations 1998 (LOLER)
- The Control of Substances Hazardous to Health Regulations 2002
- The Personal Protective Equipment at Work Regulations 1992
- Control of Asbestos at Work Regulations 2006
- The Confined Spaces Regulations 1997
- Control of Lead at Work Regulations 1998
- Noise at Work Regulations 1989.

Regulations of major importance

Although all the regulations are of great importance, there are four which stand out as being most able to influence the overall health, safety and welfare of site workers:

1. The Management of Health and Safety at Work Regulations 1999
2. CDM – The Construction (Design and Management) Regulations 2007
3. The Work at Height Regulations 2005
4. COSHH – The Control of Substances Hazardous to Health Regulations 2002.

The Management of Health and Safety at Work Regulations 1999
The Management Regulations deal with the assessment of risk and arrangements for and competence in the measures needed to protect individuals and prevent accidents at work etc.

The impact of this legislation is to ensure that issues identified by risk assessments are dealt with by effective planning, organisation and control, and that procedures to monitor and review such arrangements are put in place.

The Management Regulations are also important because they impose obligations on employers to provide adequate health and safety training for employees and to communicate health and safety risks and the measures planned to deal with them.

The COSHH Regulations 2002

These regulations were designed to address some of the long-term health problems that can arise when working with materials. The material that is commonly quoted as being the most dangerous material used in construction is asbestos; however, there are many other commonly used materials that can cause severe health problems in the short and long term if adequate protection during use is not observed.

To comply with the COSHH regulations, designers and contractors must:

■ Assess health risks arising from hazardous substances
■ Decide what precautions are needed
■ Prevent or control exposure
■ Ensure control measures are used and maintained
■ Monitor the exposure of employees
■ Carry out appropriate health surveillance
■ Ensure employees are properly informed, trained and supervised.

COSHH applies to all substances except:

■ Asbestos and lead, which have their own regulations
■ Radioactive materials, which have their own regulations
■ Biological agents if they are not directly connected with the work and are outside the employer's control (e.g. catching a cold from a workmate)
■ Commercial chemicals that do not carry a warning label (e.g. washing-up liquid used at work is not COSHH relevant, but bleach would be).

CDM Regulations 2007

The CDM regulations spread the responsibility for health and safety on construction sites between the following parties:

■ The client
■ The designers
■ The CDM coordinator
■ The principal contractor
■ Subcontractors.

CDM is a set of management regulations dealing with the responsibilities of the parties identified above and with the documentation necessary to enable construction operations to be carried out safely. Within CDM are detailed regulations aimed at controlling the risks arising from particular construction tasks such as demolitions, excavations, vehicles and traffic movement etc. The regulations also provide specific rules regarding site welfare facilities.

The regulations state that where CDM applies, pre-construction information must be provided by the client and by designers and must be collected by the CDM coordinator in order to ensure that contractors are able to identify and manage site specific risks. Before starting work on-site, this information must be developed by the principal contractor into a construction phase (health and safety) plan. After being appointed, the principal contractor needs to develop the construction phase plan and keep it up-to-date. The Health and Safety File, which is prepared by the CDM coordinator in conjunction with the principal contractor, is a further statutory document required under CDM. This is a record of information for the client or end user, and includes details of any work which may have to be managed during maintenance, repair or renovation. This must be handed to the client on completion of all the works.

Safe systems of work

In order to ensure that construction work is undertaken in as safe a manner as is feasibly possible, a safe system of work strategies should be adopted. A safe system of work strategy should include:

- *A safety policy*
  A statement of intent setting down appropriate standards and procedures for establishing safe systems of work in an organisation.
- *Risk assessments*
  Identify hazards and associated risks to the workforce and others where a safety method statement will be required to control them.
- *Safety method statements*
  Provide details of how individual safe systems of work may be devised for particular tasks.
- *Permits to work*
  Restrict access to places of work or restrict work activities which are considered 'high risk'.
- *Safety inductions*
  General safety information provided by principal contractors to workers, contractors and visitors when first coming on to site.
- *Site rules*
  Set out the minimum standards of safety and behaviour expected on-site.
- *Tool box (task) talks*
  Provide particular safety information and a forum for exchange of views regarding specific operations or activities about to commence on-site.
- *Safety audits*
  A procedure for reviewing and appraising safety standards on-site, both generally and for specific activities where appropriate.

Designing for health and safety

'Design' is a complex process of many interrelating and interacting factors and demands, including:

- Form and appearance
- Structural stability
- Heating and ventilation
- Sound and insulation
- Access, circulation and means of escape
- Environmental impact
- Cost.

Good design requires a balance of these and other factors, but as design is an iterative process, different design decisions have to be made as each stage develops. As the design process unfolds, from the concept design stage to the detail design stage, so designers of buildings, whether new build or refurbishment, are constantly seeking technical solutions to design problems.

For instance, at the *concept design* stage, decisions are made concerning:

- Plan shape
- Plan size
- Number of storeys
- Floor to ceiling height
- The nature of the building 'envelope'.

Design choices here will not only be influenced by the size and shape of the site and any planning restrictions, but also by adjacent buildings, the need to incorporate services (in ceiling voids, for instance) and by the tone and general impact of the design required by the client. However, once decided, these decisions will have an irrevocable influence on the project which the design team and the client have to live and work with thereafter. These decisions are particularly influential on the economics of the design and they will determine the cost 'bracket' within which the building naturally falls.

During *scheme design*, designers' considerations will turn to the major components of the building such as:

- The substructure
- The type of frame
- Choice of external cladding
- Building services
- The design of the roof structure.

Consequently, designers will need to consider alternative foundation choices in the context of the prevailing geological conditions on the site and methods for the remediation of contamination. The economics of alternative structural forms (e.g. steel frame vs. *in situ* concrete) and whether to choose curtain walling or precast concrete panel cladding solutions etc. will also be considered at this stage of the design.

When the *detail design* stage is reached, design questions will include:

- Staircase construction (steel, timber or precast concrete)
- Lifts and access
- The quality and specification of items such as doors and windows
- Choice of finishes (walls, floors and ceilings)
- Internal fittings (fixed seating, reception desks, special features etc.).

In making decisions throughout the design process, the designer will be constantly balancing issues of *appearance*, *function* and *cost* consistent with the client's brief and budget. As the design develops, more information becomes available, and this may give rise to technical problems or design choices that might have an impact on the financial viability of the project. A further consideration for designers is that their decisions might give rise to hazards in the design, and this will also influence their final choice.

An example of this approach might be the choice of façade for a multi-storey building. The aesthetics of the design would clearly be an important consideration in this respect, but a design which required operatives to work at height fixing curtain walling panels from an external scaffold would not be attractive from a health and safety perspective. However, a design solution whereby cladding units could be craned into position and fixed by operatives working inside the building attached by appropriate harnesses to fixed points or running lines would be a much more proactive design solution.

This objective approach to health and safety in design is eminently sensible because specific solutions to particular problems have to be found, rather than adopting 'generic' designs which might not be project-specific or address the risks involved on a particular site.

Of course, no designer sets out purposefully to design unsafely, but this is what happens unconsciously in practice. Designers therefore need to give health and safety proper weighting in their design considerations by applying the principles of *preventative design*.

However, all design considerations must also be viewed in the context of any prevailing statutory legislation. This will include:

- Permission to build (planning)
- Building standards (building regulations or codes)
- Heat loss and insulation
- The rights of third parties (such as adjoining properties)
- Statutory requirements relating to the health and safety of those who may be involved in the construction or maintenance of the building or who may be affected by the building process, such as visitors to the site or passers-by.

Of course, different legal jurisdictions around the world have their own approaches to legislating the criminal code relating to the issue of health and safety in the design and construction of buildings, and there is no common standard.

In the USA, for instance, health and safety legislation is far more prescriptive than in the UK and Europe, where the basis for such legislation is a 'goal-setting' or objective standard. Thus, in the USA, legislation comprises detailed sets of rules which must be followed to ensure compliance, whereas in the UK the less prescriptive approach relies on designers and others working out their own 'reasonably practicable' solutions to particular problems within a general duty of care.

However, this objective or 'goal-setting' approach to health and safety legislation creates problems for building designers. They must not only find solutions to their design problems, but they must do so by identifying the hazards emanating from their design, by eliminating or reducing the ensuing risks to health and safety, and by considering appropriate control measures to protect those who may be affected. This is a much more difficult standard to achieve than following detailed sets of rules where design solutions are prescribed or set down in statutes.

For instance, under the UK Construction (Design and Management) Regulations 2007 (CDM), designers have a statutory duty to prepare designs so as to avoid the foreseeable risks which might have to be faced by those with the task of carrying out the construction work or cleaning the completed structure or using it as a workplace. Allied to this statutory duty is the higher duty imposed by the Management of Health and Safety at Work Regulations 1999 to assess the risks involved and to follow the general principles of prevention that is:

■ *Avoid* the risk
■ *Evaluate* risks that cannot be avoided
■ *Combat* risks at source
■ *Adapt* work to take account of technical progress.

The best way to achieve this is by considering the health and safety impact of design decisions right from the outset of a design and throughout its development. This will mean thinking about the risks attached to both concept-stage decisions and decisions made during scheme and detail design. The issues will be different at each stage, but the thought processes will be the same. Thought processes, alternatives considered and decisions made should be recorded to provide an audit trail of the designer's thinking.

When designers are unable to eliminate risk from their designs, UK legislation requires that this is stated in the pre-construction information which informs the contractor of residual design risks thereby enabling him to develop safe working methods during construction. A further duty is to consider the possible impact of the design on maintenance activities to be carried out when the building is completed and commissioned by the client, such as window cleaning and cleaning out roof gutters.

The example given in Table 1.2 takes the reroofing of an existing 1950s factory building as an example. In the UK, this objective thinking does not stop when the building work is complete. The CDM Regulations require completed structures to

Table 1.2 Reroofing of an existing 1950s factory.

Step	Think about	Examples
1	The design element concerned	1. Removing existing roofing 2. Removing existing rooflights 3. Installing new roof covering 4. Installing new rooflights 5. Installing new guttering
2	The hazards which could be present	1. Operatives working at height 2. Fragile rooflights 3. Asbestos cement roof sheeting 4. Roofing materials stacked at height 5. Weather/wind
3	The persons in danger	1. Roof workers 2. People working below 3. Passers-by 4. Maintenance crews
4	The likelihood of an occurrence and the severity if it happened	Judge the risk by considering the chance of falling and the extent of injury likely. This is a common occurrence where fragile roofs are concerned. Consider the chance of materials falling on people below. A simple calculation could be used to give a measure of risk or a judgement could be made as to whether the risk is high, medium or low
5	The control measures required	1. Can the design be changed to avoid the risk? 2. Could permanent edge protection be included in the design, e.g. parapet wall or permanent protected walkway? 3. Specify loadbearing liner sheets for new roof for use as a working platform 4. Specify non-fragile rooflights to replace existing 5. Specify a permanent running line system for the new roof for use by construction and maintenance workers
6	Any residual risks that need to be managed	These will be design issues that cannot be resolved leaving people at risk. They should be raised in the health and safety plan. This could mean that the contractor will have to provide edge protection and/or collective fall arrest measures such as safety nets

have a *health and safety file* which is passed on to the building client or subsequent owner of the property. This file contains information which would be indispensable to those who might be involved in the maintenance, adaption, alteration or demolition of the structure at some time during its life cycle.

Included in the health and safety file might be:

- Information on how existing hazards have been dealt with in the design (e.g. contaminated sites, asbestos in buildings, underground services)
- As-built drawings including the location of key services, pipework and wiring runs
- The structural principles underpinning the design (this could be important for future demolition)
- Any hazardous substances remaining (e.g. asbestos or lead-based paints in existing buildings).

Working next to existing properties

The undertaking of refurbishment and alteration work has long been the subject of potentially bitter disputes between adjoining landowners. The nature of construction work is such that there is inevitably a risk of noise, damage and weather penetration to adjacent buildings during the works, and as far as possible the contractor should take account of these potential risks and mitigate their effects as much as possible. The ideal scenario is that the two, or more, adjoining owners agree matters in a neighbourly fashion and that disputes are kept to a minimum. However, too often this is not the case and disputes arise.

The nature of refurbishment work is such that it takes place in existing buildings that are very often close to or joined to other existing buildings. This poses problems, not just in terms of the type of work and the technology that can be utilised, but also in terms of access to undertake the works. Simple issues such as the position of scaffolding can become very complex and potentially conflict laden if there is a need to erect scaffold on land that is not owned by the party doing the work. In practice issues such as scaffolding/access permits and over-sailing by crane jibs are often settled on the basis of a commercial deal between parties. It is for reasons such as this that the Party Wall etc Act 1996 came into force in England and Wales. This piece of legislation provides a framework that is aimed at preventing and resolving disputes in relation to works that affect party walls, boundary walls and excavations near adjacent buildings. The principles of party wall negotiation had been established for many years in London and were enforced through the London Building Acts. However, the introduction of the Party Wall etc Act extended these principles on a National scale.

Under the terms of the Act, anyone who wishes to undertake works of certain types must give adjoining owners notice of their intentions in terms of the technical aspects of the work and the proposed timing. In the case of existing party walls, this applies even if the work does not extend beyond the centre line of the wall.

Once they have been made aware of the proposed works, adjoining owners have several options: they can agree with the proposals, they can reach agreement with the person proposing the work on changes to the detail or timing of the works, or they can enter into a formal dispute; in the latter case the Act provides for mechanisms of dispute resolution. Works that are included within the scope of the Act include the following:

- Work carried out directly to an existing party wall or structure
- New building at or astride the boundary line between properties
- Excavation within 3 or 6 metres of a neighbouring building or structure, depending on the depth of the excavation or proposed foundations.

It is not the intention here to go in to the detail of the Act, rather to outline the broad principles of its application and operation.

An important feature of the legislation is that, despite the name, the Act does not relate only to party walls. It actually relates to any party structure that separates individual building units. For example, the floor between two domestic flats would be considered as a party structure.

The provisions of the Act afford building owners greater rights than they might normally have under common law, to undertake works on walls or other party structures that separate their property from that of an adjoining owner. Examples of such additional rights include:

- Demolition and rebuilding the party wall to a proper standard
- Underpinning of a party wall cutting into a wall to provide suitable end-bearing of a structural beam
- Raising the height of a party wall
- Cutting into adjoining owners' wall to insert a flashing.

The main requirement of the Act is that you must give advance notice to adjoining property owners of your intention to undertake works. Whilst there is no basis of enforcement or sanction for not doing so, the adjoining owner may seek an injunction to stop works if notice has not been served.

The party intending to undertake works is required to serve a formal notice on the adjoining building owners well in advance of commencement, setting out the intended works and the timescales. Once the scope and detail of the work have been agreed, often following significant negotiations, works will proceed.

The party undertaking the works must make provision for access and temporary weatherproofing as well as for ensuring that any damaged areas adjacent to the works are made good. All of the provisions for such items, as well as the detailed description and programme of works, will be set out and formally agreed within an 'Award'. This Award is the formal agreement between the parties on the intended work and the liabilities of the parties for individual aspects such as making good after completion. The Award will also contain a detailed photographic schedule of the condition of the building and adjoining buildings prior to commencement. This acts as a reference point when making good after the works are complete.

The application of the Act has resulted in a much more structured approach to dealing with party wall issues, and the principle of reaching a definitive agreement between parties prior to commencement of works has reduced the extent of disputes greatly.

Reflective summary

With reference to statutory control, remember:

■ This is a process only recently applied in the history of building.
■ The first set of national Building Regulations was published in 1965.
■ Part M of the Building Regulations deals specifically with adapting buildings to accommodate disability.
■ Health and safety need to be considered in great detail when building forms and systems materials are specified.
■ Major references for health and safety include:
  – The Health and Safety at Work Act 1974
  – The Management of Health and Safety at Work Regulations 1999
  – Control of Substances Hazardous to Health Regulations 2002
  – Construction (Design and Management) Regulations 2007.

Review task

What are the main implications of the requirements of Part M of the Building Regulations and how can these be difficult to achieve in refurbishment contracts?

What are the requirements of clients and designers under the CDM regulations, and how are these roles made more difficult in refurbishment projects?

In the area that you live or study, try to find out which buildings are listed and why.

Outline the difficulties for a developer when proposing works to listed buildings.

Read the Party Wall etc Act 1996 and set out the chronology of a typical party wall agreement.

The context of maintenance

After studying this chapter you should have developed an understanding of:

The reasons for and types of maintenance activity that affect buildings
The terminology associated with maintenance activity and programming
The classifications of maintenance activity
The concepts of obsolescence and refurbishment
Some examples of major refurbishment activities
The economic implications of maintenance and its programming

This chapter includes the following sections:

2.1   What is maintenance?
2.2   Building maintenance management

■ RICS Planned Building Maintenance Guidance Note 1990
■ BS 8210 British Standard guide to building maintenance

2.1 | What is maintenance?

Introduction

- After studying this section you should appreciate the importance of the correct maintenance of a building and you should be able to distinguish between the various forms of maintenance operation.
- The implications of failure to maintain and the possibilities of consequential loss should be understood.
- In addition, the sequencing and relative timing of maintenance operations should be understood.

Overview

During their lives, buildings deteriorate and become obsolete; they require maintenance, refurbishment and modernisation. As soon as they are built they begin the process of decay, and the deterioration of the fabric and services begins. The increased drive towards a sustainable built environment brings into focus the importance of building maintenance. From the moment that a building is completed and occupied there is a requirement to undertake maintenance activity to ensure that it performs at an acceptable level. The process of building maintenance is not driven simply by the need to correct defects that occur over time, but rather to attempt to avoid those defects in the first place. Hence it is almost ubiquitous now for new buildings and refurbished buildings to be provided with a maintenance manual which sets out the required maintenance regime for the key elements and components of the building. As modern buildings and their services become ever more complex, the provision of such manuals is taking on increasing importance in the building procurement and commissioning process.

Maintenance as part of organisational policy

Commercial organisations now tend to recognise the importance of proper building maintenance not simply from a technical viewpoint but also in terms of maintaining a satisfactory environment for workers.

In Section 1.2 the concept of buildings as financial assets was discussed. This idea is highly relevant to the processes associated with building maintenance, since the extent to which the building is maintained is driven by the need to protect the financial asset as well as the need to maximise building functionality. For this reason the organisation will generally seek to establish some form of strategic maintenance plan or policy to maximise the economic and technical effectiveness of the maintenance of its buildings. In establishing such a policy the organisation will typically take into account the following issues:

- Intended building lifespan or life to major change of use.
- Probable time periods to major repair activity for fabric and services. This is linked to the selection of materials and components and their individual and collective life cycles.

- Required standard of building condition. Not all buildings need to be kept in the same condition. For example, consider the difference in required standards between a back street garage and a high street bank.
- The allowable time between recognition of a required maintenance or repair item and its rectification.

Taking all of these factors into account a policy may be established that allows the organisation to plan and programme maintenance, and hence expenditure, in a way that supports its general strategic objectives.

Types of maintenance

The operations that we might broadly consider as 'maintenance' are varied and sometimes complex. For this reason the management of maintenance operations must be considered in the same strategic way as the processes of design and construction. Many buildings will have long lifespans, during which time the expenditure on maintenance may greatly exceed the expenditure associated with the initial construction of the building. In order to manage the processes associated with maintenance efficiently and effectively, items are generally considered within one of the following categories:

- *Planned maintenance* (often termed *preventive maintenance*): the items included within this category are those which are planned to take place at defined, regular intervals in order to keep the building in good order. Depending on the specific items, work may be programmed to take place frequently or infrequently. The intended or predicted lifespan of an element or component is taken into account and works are planned so as to avoid failure. This means that items may be renewed whilst they still function adequately. However, the risk of major failure and consequential damage to other components or elements must be considered and prevention through planned maintenance is generally preferable to dealing with the results of failure.
- *Reactive maintenance*: items within this category include day-to-day items of repair and breakdown of plant and machinery. In addition, there may be elements of accidental or malicious damage to the building fabric and services. It is generally accepted that the level of reactive maintenance required will reduce with increased expenditure on planned maintenance, since many breakdowns and failures will be foreseen and avoided.
- *Cyclical maintenance*: this category includes items that must be dealt with routinely and regularly to maintain the building in good condition. These are what might be considered as routine servicing of the building and may include items such as cleaning of drainage gullies, servicing of lifts and so on. This aspect of maintenance is planned, but is dealt with separately from the remainder of planned maintenance activity.
- *Refurbishment*: strictly this is not an element of maintenance, but is included here since many larger maintenance operations may include aspects of

improvement to the building or may trigger the decision to undertake a minor or major refurbishment rather than a simple repair operation.

Programming maintenance

Programming maintenance work relies on a detailed understanding of construction technology and component life cycles.

The issue of deciding at which point maintenance should be displaced by refurbishment is driven by a number of factors. Typically the building owner or user will consider the following in coming to the decision to effect a major refurbishment rather than simply maintaining the existing building:

- Does the building satisfy the user's requirements in terms of function, standards and technical needs?
- Does it satisfy current statutory requirements (e.g. disabled access, means of escape)?
- Is the cost of the maintenance operation excessive compared to the refurbishment cost?

These issues will become more significant as the building gets older, and it is generally the case that the level of expenditure on maintenance increases year on year. Figure 2.1 illustrates a typical maintenance expenditure profile for a commercial building.

Here we see the various components of maintenance expenditure with cyclical, planned and reactive maintenance running from the point of construction of the building. Additionally, major refurbishment operations are identified at key stages in the life of the building. As might be expected, the level of reactive maintenance reduces following a refurbishment and the degree of planned maintenance varies

Figure 2.1
Maintenance expenditure profile.

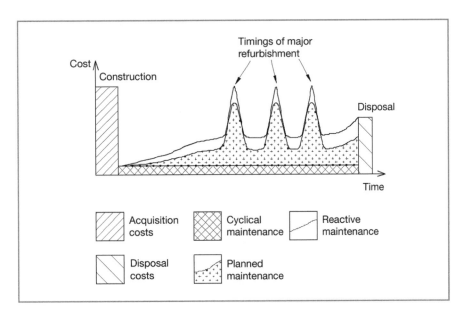

from year to year as the functional lifespan of components is reached. Planning of the maintenance programme aims to even out the expenditure profile as much as possible to allow the organisation to manage cash flow.

One significant aspect associated with the maintenance of the exterior fabric is the planning of the redecoration of the building. Since the cost of facilitating access for redecoration with the use of scaffolding or some other means is considerable, it is sensible to programme other works that require access at the same time. Hence the redecoration cycle tends to drive the exterior maintenance programme to some extent.

Typical programmes will be based around a 5–7 year cycle, using a long-term programme to attempt to forecast expenditure in broad terms. This will be supported by regular condition inspections and will feed into a shorter term programme of detailed items of repair and maintenance. This normally results in the generation of an annual maintenance programme which incorporates aspects of reactive, planned and cyclical maintenance. Each year the planned items will be reviewed to make sure that it is the correct time for their inclusion. Regular amendment of the plan will take place as works are completed, and the long-term programme may be adjusted based on this process of annual review. In addition, the programme may be adjusted in response to financial or operational pressures of the organisation, allowing management of cash flow and responding to changes in organisational policy.

Reflective summary

- The maintenance needs of buildings begin immediately after construction.
- During the life of a building, its services are likely to require far more attention in terms of maintenance and replacement than the building fabric and structure do.
- The maintenance profile of a building or facility is dictated by economic and organisational factors as well as technical factors.
- There are several broad categories of building maintenance: planned, cyclical, reactive and refurbishment.
- Maintenance costs are likely to increase over the life of the building and users may be forced to consider major refurbishment or relocation.

Review task

Generate a list of typical maintenance operations that might fall into the following categories: planned maintenance, cyclical maintenance, reactive maintenance.

Outline the key factors that will be taken into account by building users when attempting to make the decision to refurbish or relocate.

PART 1

2.2 | Building maintenance management

Introduction

- After studying this section you should have gained an understanding of how the maintenance of buildings can be managed.
- You should also be able to discuss the issues that may prompt maintenance to be undertaken and have developed a knowledge of the legislation that is relevant to building maintenance management.

Overview

Complicated service charge regimes are often put into place to ensure that landlords can recover expenditure on maintenance from tenants in commercial buildings.

Many organisations will have a clear strategy regarding the approach to be taken to ensure that their buildings are fit for use. The recognition of buildings as assets also provides a strong drive to preserve their commercial value. The effective management of property maintenance is now seen as an important element in achieving the business objectives of the building owner/occupier. Hence maintenance policies should support and integrate with the wider business policies of the organisation. Indeed, many decisions regarding the maintenance of property are made on the basis of business, rather than technical, criteria. Such decisions rely on those responsible being fully aware of the levels of funding required to ensure that maintenance requirements are met, and of the potential adverse consequences that may result from underfunding.

There is a broad range of factors that set the context and framework for maintenance activities. Strategic decisions regarding maintenance will inevitably be based upon consideration of some or all of these factors. In addition to the technical aspects of maintenance planning and programming, the broad issues that may be considered to establish the context in which maintenance operations are effected may include:

- Property management policy
- General legal issues and statutory responsibilities
- Lease issues and repair covenants etc.
- Environmental issues
- Required standards of maintenance
- Security and access
- Budgets and cost control.

The context of maintenance

As noted above, the undertaking of maintenance activity to buildings and facilities normally takes place in response to physical and functional 'triggers' affecting building performance. The extent and nature of the maintenance activity and its planning and management are also affected by those factors set out above and the decisions affecting maintenance operations may also be affected heavily by the context rather than the physical and functional triggers.

Property management policy

The status and importance of property within different organisations vary greatly. As such, the strategic approach to maintenance varies also. However, within any organisation there will be certain common elements and themes that must be considered in establishing a maintenance programme.

The nature of the core business of the building user will establish the nature and standards of the buildings in which they are required to operate. The acceptance that different users require different aspects and levels of building 'performance' is important, since maintenance standards also differ from use to use. Consideration must be given to the building's suitability in terms of its form, size, layout etc. relative to existing and projected user needs. This will assist in informing decisions regarding the potential to maintain, refurbish, upgrade or dispose of the premises in the longer term.

The extent of maintenance work required will be affected by the levels and patterns of building occupancy. These will also restrict the programming of maintenance works since there are few instances in which the undertaking of works will be possible without interfering with the day-to-day operation of the building. The pattern of use of the building may allow works to be programmed 'out of hours', although this will inevitably lead to increases in costs, as premium rates may be required by contractors for such operations.

Thus the way in which the property is managed operationally will affect the planning and programming of maintenance activity.

General legal issues and statutory responsibilities

The occupancy and management of buildings are affected by a range of statutory considerations. The purpose of this text is not to consider the legal issues surrounding property. However, it is worth noting the broad categories of legislation that will affect the scope, planning and programming of maintenance activity.

Town and country planning legislation

We considered the impact of listed building status in Chapter 1. However, the broader control systems of town and country planning legislation may also have considerable impact upon the undertaking of maintenance and refurbishment works. This is particularly the case where material alterations or extensions to the building are planned or where there will be a change of use.

Fire safety legislation

The Fire Precautions Act 1971 allows for certification of certain buildings. The layout and configuration of the building are noted as part of the certification process and any alterations that will affect the fire defence provision of the building will require consent from the local fire authority. In addition there will be a need to undertake effective maintenance of the existing fire defence installations.

Health and safety legislation

The Health and Safety at Work Act 1974 and other associated regulations (including the 'six pack' of regulations introduced in 1993) control the occupancy of buildings and the nature of the activities that are undertaken within it. In addition, they extend to the control of maintenance operations and working practices themselves. The consideration of the CDM regulations is dealt with elsewhere in this text.

Lease issues and repair covenants etc.

Most commercial and industrial buildings are occupied on a leasehold basis and are controlled by the terms of the lease, which define obligations for maintenance and repairs. The terms of the lease will define who is responsible for repair and maintenance of the building structure, fabric and services. In addition, they will define standards of maintenance works. In many cases there will be a service charge regime that allows the landlord to undertake works and recover payments from the tenants.

Environmental issues

The refurbishment and reuse of existing buildings is often a more sustainable option than new construction.

The increased visibility of environmental concerns has led to many organisations developing detailed environmental policies that set out their aspirations for dealing with the environmental management of property management and maintenance and repair activities. The degree to which this affects such activity depends largely upon the extent to which the organisation has defined an environmental policy.

Required standards of maintenance

It was noted earlier that the standards of maintenance that are applicable in some situations may differ greatly from those in others. Defining appropriate standards of maintenance is essential in maintaining an effective property management and maintenance regime. Functional or performance standards may be defined together with servicing regimes for plant and equipment.

Higher standards of maintenance will inevitably have direct cost implications, and the management of the programme should take into account the available financial resources. This requires a prioritisation process to ensure that appropriate expenditure is targeted towards the more sensitive or essential areas.

Security and access

There is an inevitable tension between the provision of ready access for the undertaking of required maintenance and repair operations and the need to maintain a secure building environment. In certain instances this is a highly significant issue, as in banks, hospitals and schools for example. The control of access and the monitoring of site activity are essential in maintaining security. In addition there may be a need to undertake a process of security vetting and of contractors' staff.

Budgets and cost control

In any organisation there will be a limit on the funding available to support maintenance and repair operations. Control and monitoring of budgets are essential parts of the effective delivery of maintenance. It is generally the case that budgets will be allocated within categories of expenditure including reactive maintenance, planned maintenance and capital improvements. The definition of such budgets relies on the ability to predict probable expenditure in advance and to manage budgets efficiently in order to deliver the maximum benefit from the available resources.

Prioritising maintenance works

The prioritisation of required works relies upon subjective judgements of the urgency of required works relative to the budgets available. The delivery of maintenance works often relies upon the ranking of requirements in order of perceived importance or sensitivity to the operations of the building and its users.

Although the ranking of required works will be based upon subjective judgements on the part of those responsible for programming and delivering the works, there will be a range of typical factors that must be taken into account. These include:

- Potential for the defects/wants of repair to get worse
- Probable pace of degradation if the defect is left unattended
- Sensitivity of the location of the item within the building
- Statutory requirements
- Potential for dealing with the work in conjunction with other programmed activities.

It is normal to prepare a detailed annual programme of required maintenance works on the basis of prior knowledge, detailed or broad-brush condition surveys, and historical data relating to repairs and defects. This programme may support a longer term planned maintenance programme, typically extending to five years and allowing a more strategic approach to be taken in planning foreseeable items of maintenance work.

Reflective summary

- Many organisations will have a clear strategy regarding the approach to be taken to ensure that their buildings are fit for use.
- The effective management of property maintenance is now seen as an important element in achieving the business objectives of the buiding's owner/occupier.
- The undertaking of maintenance activity to buildings and facilities normally takes place in response to physical and functional 'triggers' affecting building performance.
- Consideration must be given to the building's suitability in terms of its form, size, layout etc. relative to existing and projected user needs.
- The extent of maintenance work required will be affected by the levels and patterns of building occupancy.
- The occupancy and management of buildings is affected by a range of statutory considerations.
- Most commercial and industrial buildings are occupied on a leasehold basis and are controlled by the terms of the lease, which define obligations for maintenance and repairs.

Review task

Outline the legislation that needs to be considered when proposing to undertake maintenance of buildings, and discuss issues that may arise from this legislation.

Common defects encountered during construction

3 Common defects in buildings

Aims

After studying this chapter you should be able to:

> Determine the origins and mechanisms of defects
> Explain how the analysis of defects is undertaken
> Discuss the nature of defects in the substructure, walls, claddings, frames, roofs and floors of buildings
> Identify the different types of timber defect that can occur, and be able to identify the different types of defect
> Explain how damp is caused in buildings and why this can be a problem

This chapter contains the following sections:

3.1 Origins and mechanisms of defects
3.2 Analysis of defects
3.3 Substructure defects
3.4 Defects in walls, claddings and frames
3.5 Roof defects
3.6 Defects in non-timber floors
3.7 Timber defects
3.8 Dampness in walls
3.9 Flooding in buildings

Info point

- BRE Digest 251: Assessment of damage in low rise buildings
- BRE Digest 298: Low rise building foundations: the influence of trees in clay soils
- BRE Digest 299 (1993): Dry rot
- BRE Digest 307 (2003): Insect infestation of timber; Identifying damage by wood-boring insects
- BRE Digest 327 (1993): Insect infestation of timber; Insecticidal treatments against wood-boring insects
- BRE Digest 329 (2000): Cavity wall tie failure; Installing wall ties in existing construction
- BRE Digest 345: Wet rot; recognition and control
- BRE Digest 352 (1993): Underpinning
- BRE Digest 361 (1991): Why do buildings crack?
- BRE Digest 401 (1995): Cavity wall tie failure; Replacing wall ties
- Flat Roofing Association (FRA) Information Sheet No.11 Maintenance and Refurbishment
- Improving the flood performance of new buildings: Flood resilient construction: May 2007. Department for Communities and Local Government: London
- Flood Repairs Forum (2006). Repairing flooded buildings: An insurance industry guide to investigation and repair. BRE

3.1 | Origins and mechanisms of defects

Introduction

- After studying this section you should have developed an understanding of the origins of building defects.
- You should also be able to outline the processes required in order to form an understanding of the origins of building defects.
- You should be able to discuss the categories in which the effects of defects in the fabric and structure can be considered.
- You should have gained a knowledge of different building forms used in different periods of history, and be able to discuss the variety of functions that buildings need to provide in order for them to be deemed satisfactory.

Overview

The true origins of building defects can often be confused due to the wide range of potential failures and their contributing factors that may act on a typical building. However, consideration of the broad principles rather than the specific details of defects will illustrate that the origins of defects can be categorised into three broad groupings: failure of materials, design defects and workmanship error. It could also be suggested, however, that there are two further origin categories: damage by external agencies and simple wear and tear, the latter of these attempting to make a distinction between premature failure of a building element and foreseen need for replacement of an element at the end of its design life. In practice, however, the subtle distinction between an element which has failed prematurely and one which is simply 'worn out' is rather academic, since both will produce the same end result: a failed building element.

The issue of damage resulting from the actions of external agencies, on the other hand, has become the basis of considerable concern in recent years and can realistically be considered as a separate category of building defect. Occurrences such as vandalism and damage from vehicle impact, for example, may be included here.

The mechanisms by which the failures occur may of course arise from a variety of sources and may result in a wide range of potential problems for the building owner or occupier. Effective remedial action relies on having a clear understanding of defect origin. Many defects can be traced back to human error, due to simple lack of consideration in design, manufacture or assembly, corner cutting and often misconceptions on the part of the building owner or user about the level of quality for which they are paying.

Table 3.1 summarises the main classifications of defect origin with examples of how they may manifest in a typical building.

Table 3.1 Origins of defects.

Origins of defects	Examples of defects
Material failure or component failure	Deterioration of finishes such as paint Sulphate attack of ordinary Portland cement in walls and floors Metal fatigue in fixings Spalling of clay brickwork Failure of bitumen felt roofing
Workmanship failure	Joint seals DPC laps Manufacturing faults Absence or incorrect use of fixings and restraints
Design failure	Tolerance faults Material combinations and aggressive effects Difficult weatherproofing details Insufficient sizing of structural elements
External agencies	Impact damage from vehicles Vandalism Arson
Wear and tear	Natural degradation of materials

Effects of building defects

The owner and/or occupier of a building is likely to be more concerned with the immediate effects of the defect rather than its origin. However, the effective rectification of the problem will naturally rely on a full understanding of the origin and ultimate cause rather than the visible symptoms or effects. In general, the effects of defects in the fabric and structure of a building can be considered within a number of categories, as follows:

The symptoms associated with some building defects are often more serious than the underlying defect itself.

- *Weather penetration:* this category is perhaps one of the easiest to recognise, but often one of the most difficult to resolve effectively. The sources of moisture penetration which may affect a building are many and include leakages in external fabric, defective rainwater goods, and poor joint design and assembly. Diagnosis can be made difficult as a consequence of the effects of condensation and the tendency for water to track through the building fabric due to capillary action. In modern commercial and industrial buildings, leakage is often associated with ineffective seals at joints between cladding panels and differing materials.
- *Structural instability:* thankfully it is unusual for buildings to fall down due to disrepair, but the structural stability can be heavily impaired as a result of the occurrence of some defects. Where defects affect the main structural or

supporting elements of a building there is a tendency to weaken the overall structure, and this must be managed to ensure a safe building environment.

■ *Physical deterioration*: a loss of durability in the external components of a building tends to be linked with the weathering of the building fabric. Although wind and water are the prime agents in this process, the effects of ultraviolet radiation and aggressive compounds must also be considered. Internally the degradation of elements tends to arise from use or misuse by occupants and by the undertaking of processes such as manufacturing. The lack of durability tends to manifest itself in the wear and degradation of finishes.

■ *Loss of functional ability:* the functional suitability of buildings and their component parts changes with the age and use of the building, but it is undoubtedly an element which should be taken into account by the observer. The effects of a mismatch between functional performance and user requirements are dictated by both the condition and form of the building in addition to user needs. Thus one must consider the expectations and needs of the user as well as the absolute building condition.

Building age and implications on defects

As previously noted, during the course of the life of buildings a number of defects will occur. They may arise as a result of the age of the building, its structural form and type of construction, poor design, materials degradation or failure and deliberate or accidental misuse. It is important to be able to identify the typical signs of building distress and to evaluate the probable implications for building users, owners or developers. Defects arise as a consequence of a combination of factors, and therefore diagnosis is largely empirical. The first step in this process is to identify the building construction form. This relies on a broad knowledge of the typical forms associated with buildings of different ages.

Perhaps the most different building types from those of today are those dating back to the medieval period. Two forms predominated: stone construction and timber construction; few of the latter survive. Many of the major buildings of the time featured vaulted construction developed from the Roman barrel arch. However, medieval architecture differed from Roman architecture in several ways. Firstly the use of the pointed arch was introduced; secondly ribbed cross vaulting became common; and thirdly the use of arcades was developed. At this time the use of flying buttresses was developed with heavy pinnacles utilised to bend lines of thrust towards the ground. This shows a high level of empirical understanding of loads on buildings on the part of the artisan builders of the time. The forces in buildings were at this stage mainly compressive in nature, and tensile forces were problematic when working with the materials of the time.

Until the 19th century loads on buildings were still mainly compressive, as dictated by the materials that could be utilised: brick, stone and timber were the most common. However, the Industrial Revolution introduced the use of materials in quantities which were sufficient to allow for the development of new construction forms. The availability of iron of appropriate quality and in sufficient quantity was a major influence and coincided with increased demand for new building types, such as factories, warehouses, mills and railway buildings. The construction form of the time was typified by the mill building. These were usually five or six storeys high, with brick walls, timber floors and roofs and featuring large open spaces (i.e. large structural spans).

Periods of construction

Construction professionals dealing with maintenance and refurbishment must have a detailed understanding of historic and more modern construction techniques.

Most of the buildings that survive today were constructed from the Industrial Revolution onwards, and our treatment of the topic will be restricted to this period. During the period from the mid-1700s to the present day, three major periods of influence can be readily identified. By the beginning of the period, loadbearing masonry construction had effectively replaced timber-framed buildings. However, in the years up to around the mid-1800s there was a tremendous growth in population associated with the Industrial Revolution and this led to increased building production. At this stage, however, building was vernacular in its nature. From around 1850 onwards there was a change, with the organisation of labour and the introduction of new materials and the ability to transport them. Demand for buildings was high, but still focused on traditional craft-based *in situ* construction. After the Second World War the demand for building outstripped the availability of labour and materials, and the construction industry responded by introducing new designs. These reduced the quantity of materials used and increased the speed of construction, and prefabrication became common.

Each of these periods is reflected in different construction forms, each with its own specific group of common defects. The following sections provide an overview of the factors associated with the development of built form.

Industrial Revolution (mid-1700s to mid-1800s)

During the Industrial Revolution there was a massive demand for buildings of various types, many of which still survive today. The construction industry was still based on traditional labour-intensive processes, and this was continued but with the use of more materials. Statutory control was limited and the quality of some buildings constructed around this period was poor. To some extent the industrial and commercial buildings of this time were larger versions of domestic structures. They employed the same technology and were restricted by the same materials: brick, stone, timber and cast iron. In some cases the need to restrict the potential for fire to spread forced designers to attempt to remove timber from the structure as far as possible. Examples such as the many brick-built dockland warehouses that still exist show how this was achieved.

Mid-1800s to mid-1900s

By this time there was a highly developed transport infrastructure in Britain and this led to the ability to move away from vernacular architecture as material and components could be transported over reasonably large distances. Two major advances came in 1851: firstly the window and brick taxes were repealed, and secondly machinery for the manufacture of pressed bricks became available. This encouraged the larger scale use of bricks, which were now of regular size and shape. Ordinary Portland cement had been patented in 1824, but the use of concrete in buildings did not become common until the turn of the century, and even then it was still rare.

Post Second World War

Ronan Point was a system-built tower block in London that suffered a partial collapse due to a gas explosion in one of the flats.

After the war there was a deliberate move towards the prefabrication of building components and the industrialisation of the building process. High-alumina cement was utilised for a period as a way of increasing the speed of concrete construction *in situ*. Construction generally became more flexible as a consequence of the use of better quality materials and new technologies. The adoption of cavity walls now became common, windows were larger, floor spans longer and structural members lighter. In the 1960s in particular the adoption of system building became common, but failures such as the Ronan Point disaster together with other highly publicised failures damaged public confidence in these forms. However, in the industrial and commercial sector the construction of buildings has returned to the approach with many modern buildings being effectively kits of parts.

Late twentieth century

The building boom of the 1980s resulted in materials shortages and the substitution of alternative materials for the more usual options. Hence we may find ourselves dealing with building details which are less standard than it would at first appear.

The advent of the 'crinkly tin shed' school of building design has resulted in a tremendous amount of repetition in building form and consequently a database of knowledge regarding typical defects. Naturally, there has been a degree of evolution in design, but some of the early buildings still display design shortcomings, which were later eradicated. The relative newness of the building form means that many of the occurrences of defects which arise with age have yet to be fully exposed; hence there is an image of reliability which is perhaps undeserved.

Figure 3.1 summarises the development of industrial building.

Why is age important?

The long-term performance of buildings depends on a variety of factors: the location and exposure of the building, its use, the design and detailing of the structure and fabric, and the performance of the materials adopted. However, as noted earlier it is generally accepted that defects in buildings arise from one or more of the following:

Figure 3.1
Industrial building development overview.

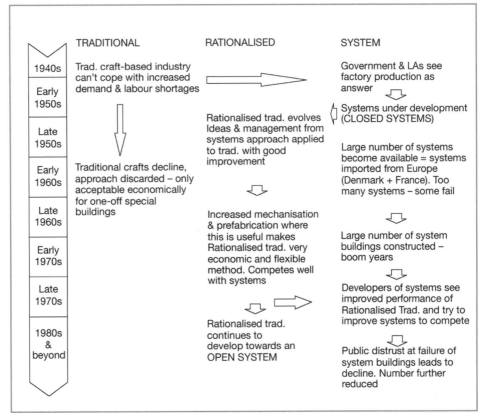

- Design
- Workmanship
- Materials
- External agencies
- Wear and tear.

The first three of these are linked closely to the age of the building and an appreciation of the differing forms will be of paramount importance in the effective identification and remediation of defects. It is worthwhile considering the performance requirements of building fabric and structure in the context of building age as an illustration of the need to be familiar with the differing approaches to construction.

Building performance requirements

Buildings are required to fulfil a variety of functions in order to be considered satisfactory, and consideration of these functions in the abstract is a useful starting point when attempting to appraise building condition and the significance of building defects. In this section we set out a number of questions associated with the performance requirements and the structural form of buildings, which is of course associated with age.

- *Structural stability*
  - Is the building framed or of loadbearing construction?
  - Are the structural members likely to be in timber, cast iron or steel?
  - What is a reasonable span for floors?
  - How are the window and door openings supported?
  - Is lateral stability affected in any way?
  - What forms are the foundations likely to take?
- *Moisture exclusion*
  - What form does the external envelope take?
  - What precautions are taken to prevent rising moisture in the floors and walls?
  - How is rainwater dealt with?
  - What happens at the junctions between components?
- *Thermal insulation*
  - How effective is thermal insulation in preventing condensation?
  - How effective is ventilation of voids?
- *Durability*
  - How durable are the materials used?
  - What was the anticipated lifespan of the building?
  - How visible are non-durable materials during inspection?

Reflective summary

- The true origins of building defects can often be confused because of the wide range of potential failures.
- The origins of defects can be categorised into three broad groupings: failure of materials, design defect and workmanship error.
- The distinction between an element that has failed prematurely and one that is simply 'worn out' is rather academic, since both will produce the same end result: a failed building element.
- Effective remedial action relies on having a clear understanding of defect origin.
- The owner and/or occupier of a building are likely to be more concerned with the immediate effects of the defect rather than its origin.
- The effects of defects in the fabric and structure of a building can be considered within a number of categories.
- Most of the buildings which survive today were constructed from the Industrial Revolution onwards.
- The long-term performance of buildings depends on a variety of factors.
- Buildings are required to fulfil a variety of functions in order to be considered satisfactory.

Review task

Outline the categories within which the effects of defects in the fabric and structure of buildings can be considered.

List the functions that buildings need to provide in order for them to be considered satisfactory.

3.2 | Analysis of defects

Introduction

■ After studying this section you should have developed a good understanding of the need for the thorough analysis of defects.
■ You should also have gained a knowledge of how surveys can be undertaken in order to assess the true origin of defects.

Overview

The analysis of defects will obviously be affected by the structural form of the building being examined, but there can be a uniform methodology for defect analysis which takes these forms into account. Survey work relies on precedent to a large degree, there being a huge information resource relating to the defects which arise in most forms of building. Such technical data must be accessed prior to the building examination in order to raise awareness of potential problems. However, a caveat should be applied here in that such empirical data may be misleading when dealing with unusual or novel details or with new materials. In examining defects in buildings three factors need to be assessed:

1. The cause of the defect
2. The effect on building performance
3. The state of activity of the defect.

This is particularly relevant to structural defects, such as cracking, which are dynamic processes that take place over time. In such situations the mechanism of the defect may be such that it is active, complete and dormant, or intermittent/cyclical.

In addition it is generally a requirement for industrial and commercial buildings that are subject to potentially complex ownership and letting agreements to establish who is responsible for the rectification of the problem. There are many cases where litigation has resulted from a disagreement relating to the 'ownership' of the defect.

Analysis methodology

There has been much research relating to the analysis and rectification of defects, and there is a vast array of publications from sources such as the Building Research Establishment and the Royal Institution of Chartered Surveyors relating to possible methodologies. This guidance generally reflects the view that assessment of a given defect or combination of defects should include some or all of the following stages:

■ *Broad brush appraisal of condition:* this will include a general familiarisation with the building form and structure, together with an assessment of the general condition of the building. This allows the specific defect and its analysis to be placed in context.

■ *Monitoring of specific defects:* it was noted earlier that some defects are dynamic in nature and it is useful to initiate a programme of monitoring (Figure 3.2 illustrates this for cracks caused by movement) to ensure that the state of defect 'activity' can be confirmed. There is little point in undertaking work to rectify a defect that is dormant.

■ *Collation of overall survey results:* bringing together information from the broad brush assessment and the monitoring exercise will allow the generation of a realistic view of the condition of a building and the possible linkages between apparently independent defects.

■ *Detailed further inspection of problem areas:* the foregoing stages will normally give indications of specific areas that need to be given more detailed attention. This may include exposure of areas of the building fabric and structure, specific specialist tests and detailed localised examination.

■ *Analysis of specific failures:* having identified the defects specifically, a process of analysis is required to identify the true cause of the problem and to allow the selection of appropriate remedial action.

With the adoption of such a structured methodology a detailed examination and analysis of defects will be possible in a wide range of buildings, relating to a wide range of building elements.

> The temptation to make snap decisions regarding the causes of building defects must be avoided. Remember that most defects are the result of a slow dynamic process.

Figure 3.2
Monitoring the movement
of cracks.

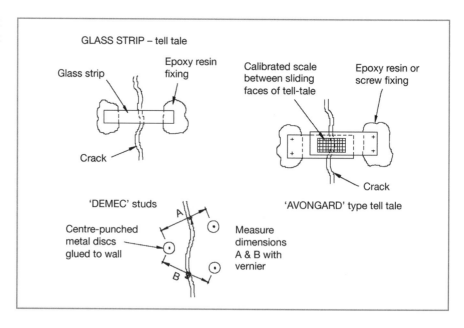

PART 2

Reflective summary

- The analysis of defects will obviously be affected by the structural form of the building being examined, but there can be a uniform methodology for defect analysis which takes these forms into account.
- There has been much research relating to the analysis and rectification of defects and there is a vast array of publications, from sources such as the Building Research Establishment and the Royal Institution of Chartered Surveyors, relating to possible methodologies.

Review task

Outline a typical structured methodology that will enable a detailed examination and analysis of defects to be undertaken.

3.3 | Substructure defects

Introduction

- After studying this section you should have developed an appreciation of how defects to buildings can occur due to failure of the substructure.
- You should also have developed a knowledge of the different conditions that can cause substructure failure, and how these conditions can be detected.

Overview

Failure of the substructure of a building is probably the biggest problem that can occur during its life, since the fundamental structural integrity of the building may be compromised. The foundations of the building form the interface with the supporting sub-strata and there must be an equilibrium established to ensure that the loads from the building are supported effectively by the ground. If the loads applied exceed the safe bearing capacity of the ground, subsidence will result. It is also important to note that subsidence may occur as a result of changing ground conditions associated with leaking drains, changes in ground conditions and so on. Hence buildings that are stable for much of their life can be affected by ground-related problems causing substructure failure later. Initial settlement of the building structure commonly occurs shortly after the construction of the building is complete, and should not cause any long-term damage provided it is uniform and not of a 'differential nature'.

There is sometimes a degree of ambiguity associated with the terms *subsidence* and *settlement*. The descriptions in this section define how these terms are interpreted within this book.

Subsidence

Subsidence occurs to buildings when the loads applied exceed the bearing capacity of the ground. This may occur as a result of insufficient foundations, poor ground strength or localised issues such as leaking drains causing erosion of the ground beneath the building. The effect of subsidence is not normally uniform throughout an entire building, and as a result it is often the case that differential movement occurs. This will normally result in one or more parts of a building moving in relation to the remainder, and the occurrence of cracking above ground may be evident as a result. The visual evidence of subsidence will inevitably be above ground, even though the origin of the defect will be at foundation level or below.

Subsidence tends to affect areas or zones of buildings, and as such the differential movement results in the generation of distinctive cracking in the superstructure of the building. Generally the cracking will be present in a recognisable pattern rather than as individual isolated cracks, although in the early stages of movement this may not be the case. The pattern of cracking will generally indicate the movement of a specific area of the building. This movement may take the form of simple dropping of a section; however, rotational movement is far more common. The nature of the movement will be indicated by the nature and type of cracks as well as the general pattern. Typically the following points should be borne in mind:

- Cracking may appear as shear cracking if vertical movement only is present
- Tapered cracks may be visible if rotation has occurred
- Tapered cracks may be wider at the top than at the bottom, depending on the mode of building failure.

Cracks will try to find the easiest route through a building, and as such they often pass between openings such as windows and doors. This may give a slightly unusual cracking pattern, but it must be identified as part of the diagnosis of the defect.

In addition to cracking of the walls, internally and externally, there may be other clues to the action of subsidence, such as:

- Upsetting of roof and coverings
- Movement between cladding and brickwork
- Settlement of surrounding areas such as paving and hard-standing
- Internal cracks at ceiling/wall junctions on settled walls
- Dropped heads to openings
- Door and window openings out of square
- Twisting of walls due to rotation of foundations.

In addition to the physical indications of movement in the building structure there is a range of factors that can also be used to attempt to identify the cause of the subsidence.

The position of drains and gullies can provide valuable information and it is often the case that subsidence is a consequence of leaking drains. In addition, the position and type of trees in the vicinity of the building can be significant. This is particularly so in areas of shrinkable clay soils, where the removal of water from the ground by trees can have a dramatic effect on the building by causing ground shrinkage and consequent subsidence of the building.

The occurrence of cracking in the walls of a building may be a result of subsidence; alternatively, there may be other causes with origins above the ground. In cases of subsidence one of the key identifying factors is that the cracking may pass through the damp-proof course (DPC) level into the ground.

Heave

Heave results from the expansion of the ground beneath the building, and its symptoms are the same as those of subsidence, but in reverse. As with subsidence, the cracking will pass through the DPC level and the cracking will be tapered, indicating a rotational movement of part of the building.

The occurrence of heave is associated with the creation of a force which acts to lift the building or part of it, causing movement and consequent cracking. This may arise in a range of situations and is often associated with areas of shrinkable clay soil, where the moisture content may vary, causing swelling of the ground. This is particularly common where trees have recently been felled and have ceased to draw moisture from the ground, and is why compressible layers are placed beneath ground beams when building in clay soils.

Figure 3.3 summarises the effects of defects below ground level.

PART 2

Figure 3.3
Defects below ground level.

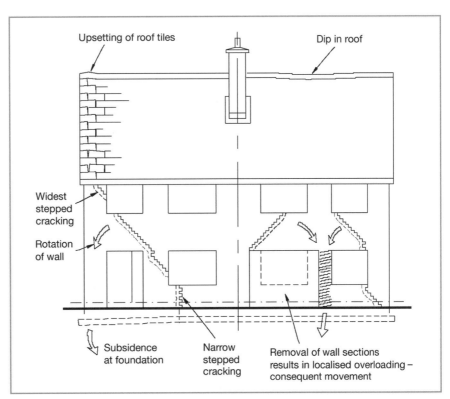

Initial settlement of new structures

All new structures settle a little in the period following construction, although this should be limited to slight movement over a fixed period and the building should not continue to move. Normally this is not a major concern; however, when dealing with the refurbishment and extension of existing buildings we must take this process into consideration. When older buildings are extended, the new section of the structure will undergo this process of initial settlement, whilst the older section will not. Hence if we attempt to tie the new section to the old there will be the risk of differential movement, and cracking in the form of vertical shear cracking will appear at the junction. For this reason connections of new sections to older buildings tend to be based on the creation of a junction that allows for such movement to take place without the appearance of cracking.

Reflective summary

- Subsidence occurs to buildings when the loads applied exceed the bearing capacity of the ground.
- The effect of subsidence is not normally uniform throughout an entire building.
- Generally subsidence cracking will be present in a recognisable pattern rather than as individual isolated cracks.
- Heave results from the expansion of the ground beneath the building, and its symptoms are the same as those of subsidence, but in reverse.
- All new structures settle a little in the period following construction, although this should be limited to slight movement over a fixed period and the building should not continue to move.

Review task

Compile a matrix to compare the causes, effects and identification of subsidence and heave.

3.4 | Defects in walls, claddings and frames

Introduction

- After studying this section you should have developed an understanding of the main causes of failure in walls, claddings and frames.
- You should be able to explain how the effects of defects in walls, claddings and frames manifest themselves and can be detected.
- You should be able to list the causes of movement in walls and be able to explain the different types of cracks that can occur.
- You should also have developed a knowledge of how the failure of supporting beams, lintels and wall ties can occur.

Overview

The wide variety of structural forms adopted for traditional, industrial and commercial buildings makes it difficult to generalise about walling and cladding types. However, there are some general principles that can be applied across a range of construction forms. The vast range of cladding and walling options makes it impossible to consider each variant individually, but we can illustrate the generic issues associated with the majority of options by reference to two common forms of walling and cladding: traditional masonry walling, as used in housing, older commercial buildings and industrial buildings; and cladding, as used in high-rise framed buildings.

Defects in masonry walls

Due to the vast range of construction details used in modern buildings, it would be impossible to discuss all in detail. However, the general principles and characteristics of different types of cracking are largely common to all materials.

The majority of defects afflicting masonry walls are associated with some form of cracking. They may arise from a variety of sources, but most involve a degree of structural movement. The consequences of settlement and subsidence are visible through cracking of walls above ground, and there is a temptation to assume that all cracking is associated with foundation defects. However, this is not necessarily the case, and a wide range of cracking-related defects are associated with causes that originate above ground. Table 3.2 sets out some of the principal causes of movement in masonry walls.

As noted, one of the major defects apparent in masonry construction is the occurrence of cracking in various forms. The presence of cracking does not necessarily mean the presence of a serious structural defect, and should be treated as one symptom of the defect along with other items of diagnostic evidence. There is a tendency to panic a little when confronted with cracking in a building; however, effective monitoring may be all that is required in the short term.

The pattern and form of cracks can indicate a great deal about the nature of movement which has caused it. Typical forms of cracking found in buildings are described below.

Tensile cracking

In this case the typical characteristics will be that the bed (horizontal) joints in the masonry will be level, with a crack opening vertically indicative of general tensile movement in the building fabric (Figure 3.4). Such cracks are often associated with thermal movement of the fabric causing shrinkage.

Figure 3.4
Tensile cracking.

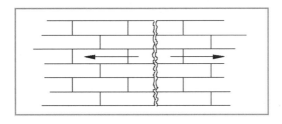

PART 2

Table 3.2 Principal causes of movement.

Cause	Effect	Duration	Comment
Temperature variation	Expansion and contraction of materials with changes in temperature	Cyclical; materials expand and contract with changes in temperature	Most building materials are affected, including brick, concrete and metal cladding
Moisture variation			
Drying	Shrinkage	This normally occurs in materials that are associated with 'wet' trades, such as plastering, and takes place shortly after construction	Materials that involve moisture in their application tend to suffer, including mortar, concrete and plasterwork
Wetting	Expansion	Materials that are delivered and used in a dry state will undergo initial moisture absorption after construction	Materials affected are those that are produced using a drying or firing process, such as clay bricks and fired ceramic products
Wetting and drying in combination	Expansion and contraction	Seasonal changes will cause a cyclical process of wetting and drying	The seasonal variation in moisture levels may cause shrinkage and swelling of clay soils. In addition, there will be degradation of materials such as brickwork, blockwork and unprotected timber
Freeze–thaw action	Expansion of cracked areas	The expansion of water as it freezes will exert repeated pressure on fissures in some materials	The expanding water/ice will cause spalling of masonry and renders and particularly affects highly porous materials
Loading			
Structural overloading	Deflection and distortion of members and elements	This depends on the nature of loading	Beams and lintels will suffer from bending if overloaded. Sections of wall may buckle or suffer from crushing
Chemical changes			
Corrosion	Expansion and lamination	Continuous	Metals and metal components, such as wall ties and other fixings, are affected
Sulphate attack	Expansion and friability	Continuous	This process affects ordinary Portland cement and hydraulic lime products, such as cement mortar and concrete, with a resultant loss of strength

Compressive cracking

Here the bed joints will also be level, but the crack will be fully closed and its edges may be subject to spalling/crumbling of the material under crushing action (Figure 3.5). These cracks are often associated with the expansion of materials due to thermal and moisture movement or with the failure to accommodate differential movement between two materials, as in the case of timber-framed housing or the occurrence of creep in concrete frames.

Figure 3.5
Compressive cracking.

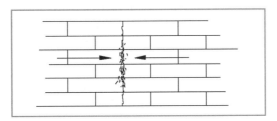

Shear cracking

Here the bed joints will be displaced or uneven, with vertical or diagonal movement evident between the sides of the crack (Figure 3.6). The crack width will generally be narrow and may show signs of a 'tearing action' as the materials on either side of the crack shift vertically in relation to each other. These cracks are often associated with overloading of localised areas, movement caused by expansion of encased elements, or differential settlement of new and old structures, as previously described.

Figure 3.6
Shear cracking.

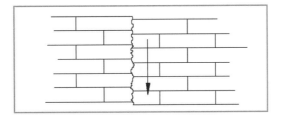

Tapered cracking

In this situation the crack width will be wider to one end and the line of the crack will often take a diagonal pattern (Figure 3.7). In traditional loadbearing walls the cracks will often extend beyond DPC level. These cracks are indicative of rotational movement to sections of a building, which may be caused by settlement, heave or deflection of supporting lintels or beams.

Figure 3.7
Tapered cracking.

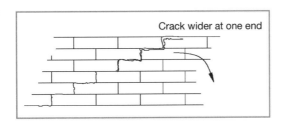

Crack wider at one end

PART 2

Defects in masonry walls arising from problems above ground

Dimensional instability

As discussed earlier, most building materials are subject to dimensional variations under the effects of changes in ambient thermal and moisture levels. The accommodation of such variations is essential if stresses are to be avoided within the material and within the overall building structure and fabric. The expansion and contraction of building components can result not only in defects occurring within the affected component but also at the junctions with other materials, where the effects of differential movement can be substantial. Table 3.3 gives an indication of the levels of dimensional variation which may be experienced in typical building materials used in external wall construction.

Movements experienced within masonry external wall and cladding components are likely to result in the occurrence of cracking due to shrinking or expansion, or of buckling due to constrained expansion of localised materials. Such cracking is most likely to manifest itself in the following locations:

- At corners of long wall sections
- On short return sections
- On long terraces or wall sections between weak spots such as window/door openings
- At junctions of different materials
- At junctions between infill panels and structural frames.

Where such cracking occurs it will generally take the form of vertical tensile or compressive cracking passing through bricks and mortar joints. It may also be visible at the junctions of infill panels and frames or at junctions between different materials, such as rendered panels. There should not be any evidence of vertical movement and the cracks are unlikely to pass through the line of the DPC.

Table 3.3 Dimensional variation.

Material	Moisture-induced change in dimension (mm per 10 m)	Temperature-induced change In dimension (mm per 10 m)
Clay bricks	2	4
Calcium silicate bricks	10	8
Steel cladding	0	8
Pre-cast concrete	10	6
Dense concrete block	12	8

Overloading of wall sections

Most modern industrial and commercial buildings which feature masonry walling or cladding have been subject to calculation of structural and non-structural brick/blockwork; hence overloading tends to be unusual. However, in older buildings, and particularly in older residential buildings, wall sizes were based upon empirical evidence and past experience, and calculation for domestic-scale construction is even now uncommon. The Building Regulations dictate the normal standards, but in older buildings these are somewhat lower than might be required today. Although such walls are generally satisfactory, certain conditions can result in cracking as a consequence of the overloading of localised sections. This tends to occur where narrow piers or isolated wall sections exist and where the loads from above are concentrated into a relatively confined area.

Concentration of the loads onto an area of insufficient strength has two significant effects:

■ Buckling of the narrow section
■ Vertical movement of the overloaded section.

Slender sections of brickwork may fail by buckling, causing collapse of the area, or they may suffer from vertical movement. If vertical movement is evident it will normally be associated with shear cracking through the masonry and passing between openings such as windows. Where buckling is evident the section will display a visible bulge and attention to address the problem should be considered as a matter of urgency.

This defect is particularly relevant to refurbishment and alteration works, since the alteration of the building layout often results in the transfer of loads onto areas of the structure that were previously loaded differently. Hence areas that were previously stable may suffer as we alter the building form. It is particularly important to bear this in mind when undertaking alterations such as the formation of new openings or the demolition of sections that may be acting as buttresses to adjacent wall sections.

Figure 3.8 shows a deformed wall section that has resulted from the absence of sufficient lateral restraint in the building structure (see Figure 3.9). Such deformation is normally arrested by the use of steel restraint fixings tying the external wall to internal floors or walls.

Failure of supporting beams and lintels

The nature of residential, commercial and industrial buildings is such that a wide variety of structural forms are prevalent. In framed construction there is little likelihood of defects associated with failed beams and supports occurring as the frames are based upon detailed calculations. However, in the case of large sheds that have traditional masonry used as cladding to the lower sections, and in more traditional loadbearing structures, such failures are common. The large sizes of openings in industrial buildings mean that any such failure can be serious in

Figure 3.8
A deformed wall section.

nature, since significant bending moments are experienced within the beams over openings to loading bays etc. In domestic-scale construction the effects are likely to be far less significant, although the principles are exactly the same. The failure of a supporting beam or lintel above an opening in a masonry wall will show a range of typical characteristics (Figure 3.10):

- Vertical movement may be visible, resulting in cracks which may adopt a distinctive triangular pattern
- Vertical cracking between openings located in a vertical line is common
- Bowing of lintels or (in older properties) dropping of brick arches may be present.

Beams and lintels may fail as a result of a variety of causes, and a detailed understanding of the wider structural form of the building and its components is essential in drawing conclusions regarding causes, effects and remedies. Typical causes associated with these failures include overloading of the beam, corrosion of metal beams resulting in loss of strength, and timber decay affecting older forms of building. Settlement of the building may also cause removal of support to the end of the beam. This effectively creates referred failure of the lintel or beam.

In older buildings, in which softer lime mortars were used, it is often the case that although movement has taken place there will be no visible cracking, as the soft mortar is capable of accommodating the movement.

Failure of cavity wall ties

Cavity walls have been used in construction in Britain as a standard form since the 1900s, with several types of tie utilised to link the inner and outer leaves of loadbearing walls, or more recently to restrain cladding to inner linings, as in

Figure 3.9
Overloading of walls.

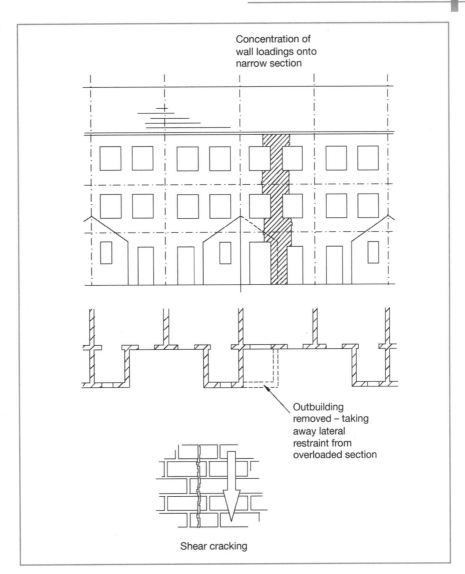

Concentration of
wall loadings onto
narrow section

Outbuilding
removed – taking
away lateral
restraint from
overloaded section

Shear cracking

timber-framed house construction. The function of the tie is to ensure that two narrow sections act as a single unit, with the rigidity of a wide structure. In early domestic structures ties were commonly formed from timber, slate, brick headers, cast iron and wrought iron, although newer buildings tend to be provided with galvanised or stainless steel ties or fixings. Mild steel replaced the early domestic forms, but in its early use was not sufficiently protected from corrosion, and hence lamination occurred, giving rise to cracking.

More modern forms of tie are normally galvanised steel wire or similar, and do not give rise to the same level of expansion and cracking as the older heavier steel ties, which were much heavier in section. Hence the first sign of their failure may be bowing of wall sections, leading to instability, rather than the typical pattern of cracking associated with wall tie failure. A common defect in framed buildings,

Figure 3.10
Failure of lintels.

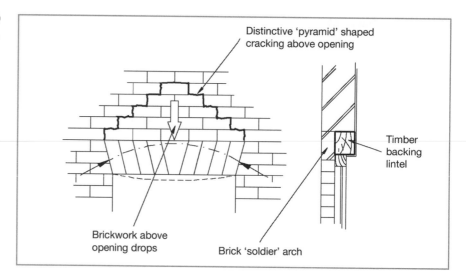

Distinctive 'pyramid' shaped cracking above opening

Timber backing lintel

Brickwork above opening drops

Brick 'soldier' arch

notably those constructed in the 1960s and 1970s, is the failure to include sufficient numbers of ties or restraint fixings during construction. This stems largely from poor workmanship caused by lack of familiarity and the need to construct quickly, together with the failure of designers to provide for sufficient tolerance for inaccuracies in the construction process.

The corrosion of older types of tie results in lamination and expansion of the cross-sectional area of the tie in the exposed outer leaf of the wall. Exposed elevations tend to suffer most from the effects of corrosion. The result of the expansion of the tie is that the bed joints of the external leaf are forced open, giving rise to distinctive horizontal cracks at regular intervals. The extent of the cracking tends to be greater at higher points on the wall due to the decreasing weight acting on the corroded tie.

Figure 3.11
A twisted and corroded steel wall tie.

However, cracking of this type results from the lamination of relatively large sections of metal. Newer ties are far more slender, and the effect of corrosion is generally a wasting of the tie, resulting in a reduction in strength and eventual loss of restraint. In this situation internal inspection of the cavity using an endoscope is the only real way of assessing the condition of the ties. In extreme cases the wall will show bulging, and this may result in a different pattern of cracking altogether.

Figure 3.11 shows a twisted steel wall tie that is suffering from corrosion to the section embedded in the outer leaf.

Sulphate attack in ordinary Portland cement

Ordinary Portland cement (OPC) is a relatively recent invention. Prior to its use, lime-based mortars were used.

Materials containing ordinary Portland cement can suffer from aggressive action of soluble sulphates, which affect the structure of the material. The active element of ordinary Portland cement is tricalcium aluminate, which reacts with the sulphates to form calcium sulpho-aluminate, resulting in expansion of the material and a loss of strength. In addition to a general loss of strength, sulphate attack results in friability of the material. Hence the visual signs will include expansion of mortar joints above ground level with visible crumbling of the mortar in affected areas. Quite often this is accompanied by a visible overhang at DPC level (Figure 3.12) as a result of the fact that the brickwork above DPC level is unrestrained and expansion takes place readily, whilst that below is effectively restrained by the ground.

Another common area where sulphate attack shows visible evidence is in masonry chimney stacks (Figure 3.13), where sulphurous flue gases affect the mortar joints. In this situation the more exposed side of the stack will often be colder, and condensation will affect that side to a greater extent; hence the typical curved appearance of stacks as the mortar joints on the exposed side expand while those on the sheltered side remain dormant.

It should be noted that the effects of sulphate attack are irreversible, and they can only be stopped by arresting the source of water which carries the soluble sulphates. In extreme cases the only remedy is rebuilding of the affected areas.

Figure 3.12
Sulphate attack in walls.

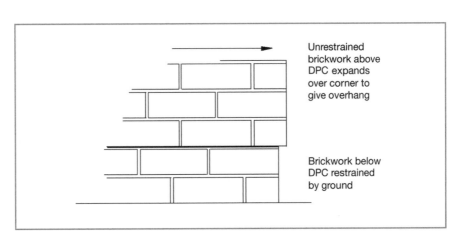

Unrestrained brickwork above DPC expands over corner to give overhang

Brickwork below DPC restrained by ground

PART 2

Figure 3.13
Sulphate attack on chimneys.

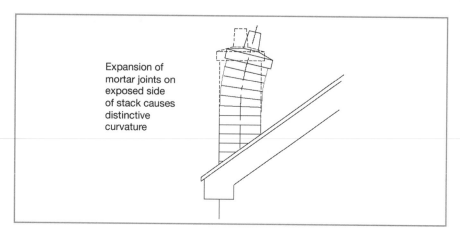

Expansion of mortar joints on exposed side of stack causes distinctive curvature

Defects in claddings

As with other building elements, the defects that occur in claddings to buildings arise from failure of materials, workmanship or design, and often result in weather penetration in combination with other symptoms. One of the most common causes of defects in cladding is the occurrence of differential movement between elements of the structure and fabric. Differential movement (Figures 3.14 and 3.15) may take a variety of forms and give rise to a number of symptoms, including cracking, bulging and distortion (Figure 3.16), misalignment of panels and moisture penetration.

Figure 3.14
Differential movement between cladding and structure.

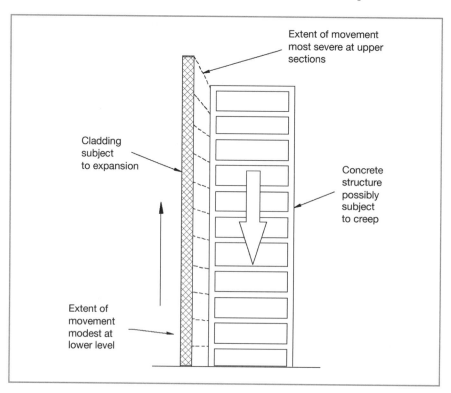

Extent of movement most severe at upper sections

Cladding subject to expansion

Concrete structure possibly subject to creep

Extent of movement modest at lower level

Figure 3.15
Provision for movement
in fixings.

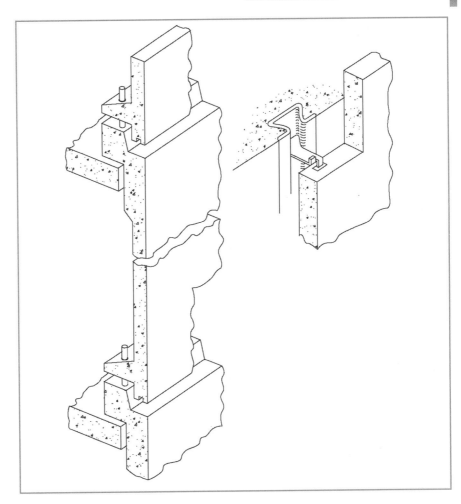

The vast range of possible cladding systems makes it impossible to consider them all here. However, some of the broad principles are set out in BRE Digest 217.

BRE Digest 217 (Table 1) gives advice relating to the frequency of inspection of claddings in a variety of situations (see Table 3.4).

Table 2 of the BRE Digest considers the indication and significance of defects within four categories:

■ Walls of clay brickwork
■ Stone of pre-cast concrete cladding
■ Sealant-filled joints
■ Light cladding and curtain walling.

Key factors in the occurrence of defects in claddings will inevitably vary from situation to situation. However, the following is a good checklist for evidence of significant defects:

■ Thermal movement
■ Moisture movement

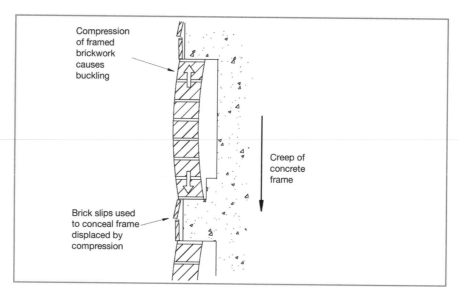

Figure 3.16
Brick cladding on concrete frames.

- Sulphate attack
- Compression of cladding
- Corrosion of reinforcement
- Rain penetration
- Condensation
- Cracking.

The lighter forms of cladding, such as rain screening and curtain walling, will also tend to suffer from a series of defects, often related to the tendency to move relative to the main structure. This is commonly manifested in the form of distortion from the vertical plane. BRE suggest that there is a need to accommodate up to 75 mm distortion from the vertical during the assembly process. This is further exacerbated by the distortions between claddings and frames which occur in use

Table 3.4 Recommended frequency of inspection of claddings (from BRE Digest 217).

Situation	Frequency of inspection
New buildings or refurbished old buildings	Every year for first 5 years
Parts of buildings at risk of vandalism	Annually
Severely exposed parts	Every 3 years
Parts over public areas	Every 3 years and after severe gales, snowfall or other extreme weather conditions
Other buildings	Every 5 years (or when repainting)
Where defects are known to exist	As appropriate

as a result of thermal and other stresses. The specialist nature of these proprietary forms is such that they may form the basis of a complete text in their own right; hence they are not considered here.

Defects in frames

As previously noted, it is uncommon for structural frames, whether long span or high rise, to suffer from serious defects in their own right. Most defects associated with frames are related to the interface between the frame and cladding, often associated with deterioration or design faults affecting fixings. However, some defects do occur due to insufficient allowance for tolerances within the system.

Assembly problems

It is unlikely that the casual observer will notice assembly faults in a frame, since they will inevitably be concealed by the fabric of the building. However, the requirement for a degree of tolerance in the position of connections and fixings can sometimes lead to misaligned connections and difficulties in ensuring that all fixings are secure. This is thankfully rare in the actual frame connections, but the fixings that secure claddings etc. are likely to suffer more commonly from this.

Creep in concrete frames

In the period after construction, concrete frames undergo a process of 'relaxation' which results in 'creep' or shortening of the frame. As such, any claddings or other elements fixed to the frame should be designed and connected so as to accommodate this slow and modest degree of dimensional change in the frame. Failure to do so will result in compression cracking in some elements and the creation of compressive stresses in claddings etc.

Corrosion of reinforcement

The construction of frames using reinforced concrete is well established and generally offers a durable, inert design solution with the added advantage of being inherently fire-resistant. However, in some situations there can be major problems associated with the deterioration and degradation of the frame as a consequence of corrosion of the reinforcing steel. This can result in visible staining on the surface of the concrete, together with spalling and cracking of the concrete due to the lamination and expansion of the steel within the concrete.

In normal situations the concrete covering the steel reinforcement provides an effective barrier against moisture and inhibits corrosion due to the alkaline nature of the concrete. Corrosion can result from failure of the physical barrier, thus allowing moisture penetration. This can occur where the concrete is highly porous, where there is some physical damage or where there is simply insufficient thickness of concrete to provide effective cover to the steel. Alternatively, corrosion may result from a change in the level of alkalinity within the concrete, reducing the degree of protection to the steel. Such chemical changes can result from the long-term penetration of chemicals from the local environment. One of the most common causes, however, is the inclusion of chemicals during production of the concrete. For example, calcium chloride was commonly added to concrete during mixing as a mechanism to promote rapid strength gain. However, this is no longer recommended.

Perhaps the most common cause of corrosion to reinforcement is carbonation of the concrete resulting from the action of carbon dioxide in the air. The carbon dioxide acts with moisture in the pores of the concrete to create carbonates that are acidic in nature. The chemical changes in the concrete result in a gradual loss of alkalinity and the carbonation penetrates deeper into the concrete with time. The quality and density of the concrete have a significant effect on the propensity to suffer from carbonation; weaker concrete is more prone to the problem. Once the carbonation reaches the steel within the concrete the alkaline protection will be lost and the steel begins to corrode, causing spalling as the lamination and expansion of the steel force the surrounding concrete away.

Reflective summary

- The vast range of cladding and walling options makes it impossible to consider each variant individually.
- The majority of defects afflicting masonry walls are associated with some form of cracking.
- Typical forms of cracking found in buildings are tensile cracking, compressive cracking, shear cracking and tapered cracking.
- Most building materials are subject to dimensional variations under the effects of changes in ambient thermal and moisture levels.
- The movements experienced within masonry external wall and cladding components are likely to result in the occurrence of cracking due to shrinking or expansion, or of buckling due to constrained expansion of localised materials.
- In framed construction there is little likelihood of defects associated with failed beams and supports occurring, as the frames are based upon detailed calculations.
- In older buildings, in which softer lime mortars were used, it is often the case that although movement has taken place there will be no visible cracking, as the soft mortar is capable of accommodating the movement.
- A common defect in framed buildings is the failure to include sufficient numbers of ties or restraint fixings during construction.
- Materials containing ordinary Portland cement can suffer from aggressive action of soluble sulphates, which affect the structure of the material.
- One of the most common causes of defects in cladding is the occurrence of differential movement between elements of the structure and fabric.

Review task

Outline the main reasons for defects occurring in building frames.

List the causes of the following in masonry walls:

(a) tensile cracking
(b) compressive cracking
(c) shear cracking
(d) tapered cracking.

3.5 | Roof defects

Introduction

- After studying this section you should have developed an understanding of the reasons for the occurrence of defects in roofs, specifically flat roofs and low pitch roofs.
- You should also have attained a knowledge of how the design of these types of roof has evolved in order to reduce the number of defects that can occur.

Overview

The nature of this text is such that it deals with a wide variety of different building types and forms. As such, some of the technology that we have considered has been generic and capable of application to many different building types. In the case of roofs, however, we shall focus primarily on the technologies associated with larger building forms, since many of the defects associated with roofs of domestic buildings are actually dealt with in the section on timber defects. The roofs utilised in the construction of commercial and industrial premises tend to be restricted to two types: flat roofs and low pitch roof forms. Within these categories there is a wide variety of defects, of which we consider only some examples of typical problems.

Flat roof defects

Modern flat roof design has largely been successful in eradicating the large-scale flat roof failures of the past. The advent of high-tensile elastomeric felts has helped to reduce the number of defects arising from failure of membranes due to thermal stresses and differential movement between roof structure and covering. Some of the modern thinner alternative membranes may be susceptible to damage from external agencies, although this is largely linked with location and access to the roof area.

When considering the possible problems associated with roofs generally and flat roofs in particular, it is important to remember that many apparent roof leaks have actually been noted to originate from causes other than failure of the roof covering. For example, many cases of apparent 'roof leak' are associated with problems at junctions and within external walls, and may be due to factors such as (see Figure 3.17):

- Defective flashings
- Absence or failure of cavity trays.

In addition, faults related to poor design and workmanship standards may occur and may afflict all flat roof types, irrespective of specific constructional details. These are described below.

Figure 3.17
Moisture penetration in flat roofs.

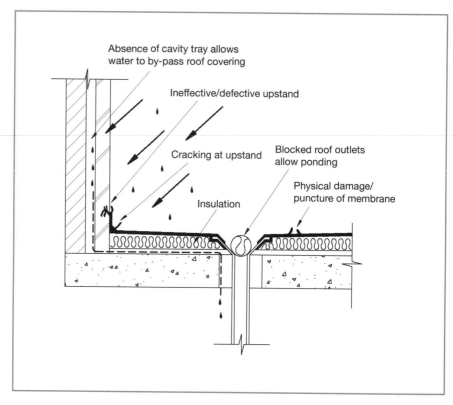

Ponding due to insufficient gradient

All flat roofs require a minimum level of inclination or fall to allow rainwater to run off the surface. If there is insufficient fall it will likely result in the occurrence of ponding on the surface of the roof. This may be due to poor design, in which insufficient allowance is made for the provision of fall to promote run-off of the rainwater. Alternatively, it may result from deterioration of the roof deck, resulting in localised deflection allowing the water to pond in a depression in the roof surface. In roofs constructed of timber, as in most domestic flat roofs, there is also the risk of sagging of the roof joists, resulting in a loss of fall and the potential for ponding.

Poor edge detailing

The edges of the roof covering will provide a danger point for moisture penetration since the impervious covering will need to be sealed effectively at this location. Many instances of flat roof leakage actually result from moisture passing behind a poorly detailed upstand or edge detail. An example of this is the situation in which asphalt upstands are prone to sagging if there is insufficient key to provide support. The inclusion of expanded metal lathing to aid in providing a key can avoid this defect.

Differential movement

It was noted in a previous section that all materials change their physical size with changes in temperature. This is a particular problem in the construction of flat roofs, where the differential movement of the supporting structure, the deck and the impervious membrane can cause problems. In older forms of covering, such as bitumen felt, the membrane was quite inflexible. Hence when the substructure of the roof expands due to thermal changes the membrane is forced to stretch. Upon cooling, the substructure will shrink back to its original size, but the membrane is likely to ripple rather than shrinking back to its original state. This is a process that occurs very slowly, but which can result in the creasing and failure of the roof membrane. The problem is exacerbated as the membrane gets older and as a result of the action of ultraviolet radiation becomes brittle. At this point cracking may occur, and moisture penetration will be the consequence.

The technology of flat roof coverings has seen great advances in recent years and the rather poor reputation for durability is becoming less deserved.

Ultraviolet radiation

Bitumen and asphalt roof coverings suffer from the loss of volatile elements under the action of ultraviolet radiation in sunlight. Hence the surface of flat roofs is normally protected by the application of a layer of mineral chippings or the provision of a painted reflective coating. The treatment of upstands and gutters etc. should be carefully considered, as these are more difficult to protect. As such, they tend to suffer more from degradation and potential failure.

Entrapped moisture

Where flat roofs are formed with an *in situ* concrete deck there is the possibility that moisture used in the formation of the deck will remain trapped after the impervious layer has been applied to create a waterproof flat roof. The effect of this is that, under certain circumstances, as the roof is subject to solar radiation the water will vaporise and expand beneath the membrane, causing blistering and possible failure of the covering. This is avoided by the use of vents in concete decks to allow the deck to liberate the water vapour without compromising the integrity of the covering. In timber roofs the eaves and verges provide this ventilation.

Defects in industrial pitched roofs

When first introduced, profiled metal roofing tended to adopt site assembly details, which combined traditional construction materials such as timber spacers and battens with the external cladding sheets and internal linings. Fixing battens tended to be formed of site-cut timber, with internal linings being formed from plasterboard or some other lining board material.

Later details improved the situation, with the incorporation of steel spacers to replace the timber battens and the inclusion of profiled steel lining panels. Interior lining sheets were arranged in such a manner as to facilitate the drainage of minor leakages and condensate to the roof guttering, thus giving the appearance of weather tightness.

Defects in this form of roof cladding tend to be associated with moisture penetration or more commonly condensation, resulting in moisture presence internally. The provision of an inner lining prevents accurate assessment of the cause, location and extent of such problems.

The durability of metal sheet cladding is generally good, and it is unlikely that serious defects will occur to the body of the cladding within its design life. However, deterioration of the surface finishes, which may be of a variety of types, is common. Causes of such failure include excessive heat build-up resulting in lifting of applied paint finishes – surface temperatures of 80°C may be experienced on dark-coloured cladding. Such variation in temperature also has an effect on the fixings that are used to secure cladding to the purlins or cladding rails of the building. Aluminium sheeting will expand by 10–13 mm over an 8 m length when provided with light and dark finishes respectively. Hence, for significant lengths of sheeting, oversized holes or non-rigid fixings should be used if the effects of movement are to be successfully accommodated.

Some key elements to be aware of when examining sheet roofing or cladding are described below.

Cold night radiation

The large volumes of air contained within industrial buildings will be capable of maintaining relatively high levels of moisture due to the warm internal environment. In buildings where there is little in the way of insulation to the fabric, there is the potential for condensation to occur at the point of contact between the warm moist air and the cold external wall or roof cladding. This is particularly the case in situations where the temperature drops rapidly on the outside, as on clear, cold nights, and internal condensation can be significant.

Fixing failures

Where cladding is secured to timber battens, as in the older details, there is potential for the battens to shrink, causing the fixing to become loose and allow moisture to penetrate around the screw head. In newer details this is unlikely, but the repeated expansion and contraction of the cladding with thermal variation may result in elongation of the fixing hole and possible minor leakage.

Deterioration of surface finish

The application of a plastic-type protective finish to steel cladding sheets ensures durability and resistance to corrosion. However, the manufacture of the sheets relies on the application of the plastic finish as the steel is extruded in long lengths. When the sheeting is cut to appropriate length or trimmed on-site it is important that the cut end is protected to prevent corrosion. If this is not done the sheets will corrode from the end and the coating will be lifted from the metal.

Figure 3.18 summarises the typical types of damage in pitched roofs.

Figure 3.18
Defects in industrial pitched roofs.

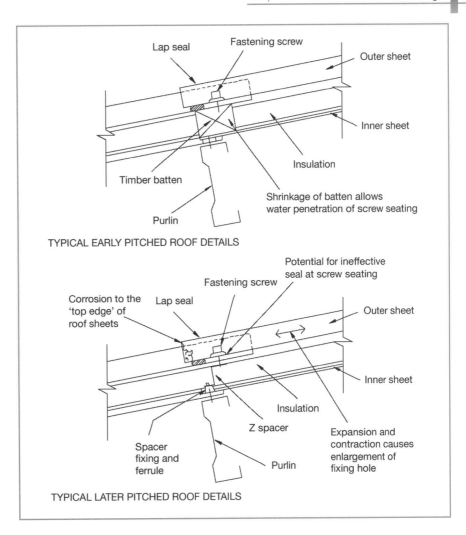

TYPICAL EARLY PITCHED ROOF DETAILS

TYPICAL LATER PITCHED ROOF DETAILS

PART 2

Reflective summary

- The roofs utilised in the construction of commercial and industrial premises tend to be restricted to two types: flat roofs and low pitch roofs.
- Modern flat roof design has been largely successful in eradicating the large-scale flat roof failures of the past.
- Many cases of apparent 'roof leak' are associated with problems at junctions and within external walls.
- When first introduced, profiled metal roofing tended to adopt site assembly details, which combined traditional construction materials such as timber spacers and battens with the external cladding sheets and internal linings.
- Defects in this form of roof cladding tend to be associated with moisture penetration, or more commonly condensation, resulting in moisture presence internally.

Review task

List the most common causes of defects in flat roofs.

3.6 | Defects in non-timber floors

Introduction

- After studying this section you should have developed an understanding of the reasons that floors fail.
- Failure in timber floors is detailed more throughly in Section 3.7, whereas this section deals mainly with failure in concrete floors.

Overview

There are numerous defects associated with floors to buildings, and the nature of the defects will vary, depending upon the construction form of the floor. The use of timber ground floors in modern buildings is quite rare, although the formation of upper floors using timber is a common feature of domestic buildings and older forms of commercial and industrial buildings. The defects associated with timber floors are largely dealt with in Section 3.7, and we shall not cover them here. Instead, we will deal with some of the more common defects in other types of floors. It is generally the case that in modern domestic buildings and in industrial and commercial situations, floors will be of ground-supported concrete or suspended concrete formation.

Typical defects in concrete floors are described below.

Surface abrasion

The surfaces of power-floated floors, most notably to industrial buildings, can suffer from degradation as a result of loads and traffic across the surface. This results in damage to the sealed surface and potentially in fragmentation of the floor adjacent to construction and movement joints.

Surface crazing

When the surface screeds of solid concrete floors are laid, it is important that the curing and drying processes are controlled to ensure that the material achieves the desired strength and composition. If the material is subject to rapid drying, the evaporation of moisture from the surface encourages the fine cement particles to migrate to the surface. This can result in a variable density in the screed and the subsequent occurrence of surface crazing. In practice this is inhibited by the use of impervious membranes laid over the floor during the curing/drying process to prevent rapid evaporation of the moisture.

Structural movement

The settlement of solid floors is not uncommon and is sometimes associated with localised heavy loads applied to the surface, poor compaction of back-fill around foundations or poor ground conditions beneath the floor. The symptoms of such movement are easily recognised, as the floor surface may suffer from deformation and cracking of the surface may result. Figure 3.19 shows an example of structural movement. The exposed section of brickwork to the centre of the photograph illustrates the degree to which the floor has suffered structural movement in this industrial building.

Figure 3.19
Structural movement of a solid floor.

PART 2

Curling of screeds

The use of cementitious screeds to provide a level and uniform wearing surface for floors is a well-established construction technique. However, it is important that the screed is laid of sufficient thickness if curling at the edges of construction sections is to be avoided. The use of thin screeds (below 50 mm) can result in curling and separation as the screed shrinks during drying.

Sulphate attack

Sulphate attack is often associated with the presence of contaminated hard core materials, clay soils, filled sites or areas in which chemical or industrial processes take place.

The effects of sulphate attack, as previously described for wall defects in Section 3.4, can also have serious effects upon concrete floors. The expansion of the concrete that results from sulphate attack can induce hogging of the restrained floor slab as it seeks to expand within the confines of the restraining external walls. This also results in cracking and crazing of the floor as it deforms and is affected by friability associated with sulphate attack.

Review task

Explain the causes of common defects in concrete floors.

3.7 | Timber defects

Introduction

- After studying this section you should have developed an understanding of the main causes of timber defects. These fall into two broad categories: rot and insect attack.
- You should be able to explain what causes the development of these forms of attack and be able to identify the different types from photographs.
- You should then be able to detail the types of remedial action that would be required in given scenarios.

Overview

Timber is an inherently durable material and is resistant to most forms of degradation if maintained in a dry condition. However, given certain combinations of circumstances, deterioration can occur. This typically takes the form of an attack by wood-rotting fungi, which can develop in the timber when moisture levels are maintained above about 22 per cent for a prolonged period or attack by wood-boring insects. Attacking of timber by wood-boring insects is often linked with high moisture levels, but this is not always the case.

The effects arising from the attack of fungi and insects range from total physical deterioration of the timber, requiring extensive treatment and replacement, to simple staining, requiring no treatment whatsoever.

The effects of wood-rotting fungi will typically give rise *inter alia* to the following effects or symptoms:

- Loss of strength, disintegration or softening of the wood
- Discolouration of the affected area
- Presence of mycelium strands and fruiting bodies
- Musty smell
- Hollow sound when struck.

Insect infestation will typically produce the following symptoms:

- Loss of strength, disintegration or softening of the wood
- Flight holes at the surface of the timber
- Bore dust in surrounding areas
- Insect frass
- Presence of larvae
- Fungal attack.

There are several forms of fungus which can affect timber in buildings, but which may not necessarily cause wood 'rot'. The presence of any fungus indicates that the conditions are damp; hence there may be a real danger of the generation of other more debilitating fungal growth. The fungal attack of timber in buildings can be split into two generic groups: wet rots and dry rot.

The effects of most wet rots are essentially similar, as are the measures adopted to eradicate them. Hence they may be treated as a general type. The effects of dry rot, by comparison, may be devastating and may require expensive and extensive remedial action, in a quite different manner from that required for the treatment of wet rots. Hence it is crucial that the correct identification, at least of dry or wet rot types is made, to ensure that the appropriate remedial action is taken.

The identifiable characteristics of different fungus types are linked to their life cycle and the stages of development which they undergo. Figure 3.20 illustrates these stages.

Wood-rotting fungi tend to develop in environments with a moisture level of above 20 per cent in the timber. Airborne spores may be present which will germinate if they are allowed access to a favourable environment, such as damp wood. Upon germination the spores develop into thin strands or hyphae, which when present in sufficient quantities create a body of mycelium. The hyphae enter the body of the wood and break down the cell walls to feed on them. At advanced stages of fungal growth a fruiting body is produced that generates further spores, which are dissipated into the air. The cycle then starts again.

These discrete stages of fungal development provide the features which may be used to aid identification of the fungal types. Other features must also be taken into account and the following is a brief summary of the main, identifiable features of fungal attack:

- *Fruiting bodies and spore dust:* the structure and appearance of the fruiting bodies of each fungus species is unique; hence identification of the fruiting body is by far the best means of identification. However, since their presence is generally linked with advanced stages of attack, or is often concealed, this is not always possible.

PART 2

Figure 3.20
Typical fungal life cycle.

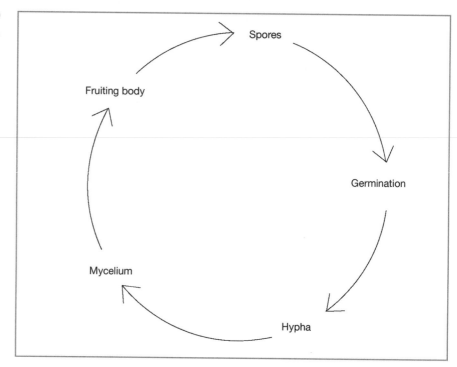

- *Hyphae and mycelium:* if identification of the fungus is not possible by examination of a fruiting body, the presence of mycelium is the next best indicator of species. As with fruiting bodies, the appearance and colour of mycelium varies widely between species.
- *Physical appearance of the wood:* the examination of the physical condition of the timber is an inherently unreliable mechanism for identification, since many features are similar for both wet and dry rots. Such features can be of help, however, in confirming the findings of other investigations.

The main fungal types can be broken down into generic families and are generally considered under the groupings of brown rots and white rots.

- *Brown rots*: These forms of fungal attack will generally cause a darkening of the wood, together with cracking along and across the grain. Several common wet rots fall into this category, as does dry rot.
- *White rots*: these forms will tend to cause the wood to become lighter in colour and to take on a fibrous appearance. Cross cracking will not occur. All forms within this category are wet rots.

In order for dry rot to develop a very specific set of environmental conditions is required. Ensuring a dry, well-ventilated environment should prevent an outbreak.

Dry rot

Within the category of brown rots is undoubtedly the most serious form of fungal attack: dry rot (see Figures 3.21 and 3.22 for examples). This is potentially the

Figure 3.21
Examples of dry rot.

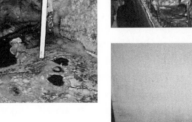

Figure 3.22
Further examples of dry rot.

most devastating of the forms of fungal attack which occur in buildings and must be controlled to prevent serious damage to the timber elements of the building fabric and structure. Typically occurring in wood which is in contact with wet brickwork, for example, and capable of growing through mortar joints etc. in search of nutrition, the fungus is sensitive to high temperatures and does not readily thrive in areas where conditions fluctuate or where exposed to the open air. The fruiting body may often be the first visible sign of an outbreak.

The recognition of dry rot is based upon the key elements of the fungal life cycle noted earlier:

- *Fruiting bodies and spores:* these are usually found at the junction of wood and walls/floors, internally. They are yellow in colour when young, but develop into a brown/red colour with age, often with a grey/white fringe. Spores appear as fine red/brown dust with profusion.
- *Hyphae and mycelium*: hyphae appear in varying thicknesses, sometimes up to 7 mm diameter, and they may show signs of brittleness when dry. Mycelium appears as cobweb type sheets in white/grey. In advanced stages this may present an appearance of cotton wool and may be tinted with yellow or purple where exposed to daylight.
- *Physical appearance of the timber:* affected timber is brown in colour and will be light in weight, giving a hollow sound when knocked. There is unlikely to be a sound skin material over the decayed areas. The surface of the wood may show signs of deep cracks along and across the grain.

Wood-boring insects

As in the case of fungal attack, the extent of damage caused by insect attack, and the level of remedial action necessary, are variable. In many cases, the presence of insect infestation does not warrant any remedial treatment, but in others it is essential that the infestation is eradicated in order to prevent extensive damage to the timber. The types of insect found in buildings will fall into one of three categories, defined by the level of treatment required to prevent deterioration of the timber, as follows:

- *Insecticidal treatment is required*: insects in this category include the common furniture beetle or 'woodworm' (*Anobium punctatum*), house longhorn beetle (*Hylotrupes bajalus*) and powder post beetle (*Lyctus brenneus*).
- *Treatment is required to control an associated fungal attack*: insects in this category include wood-boring weevils and wharf borers.
- *No treatment is required*: insects in this category include bark borers and wood wasps.

The nature and extent of damage resulting from attack by wood-boring insects will vary greatly from situation to situation. The nature of the remedial action necessary is thus also variable. The appropriate response to an insect attack can vary tremendously and it is vital to ensure correct identification of the wood-boring insect.

Identification of insect attack

Insect attack is categorised into three broad groups as defined above, according to the nature of the damaged caused. As in the case of fungal attack on timber, the identifiable characteristics of insect species are linked with their life cycle (Figure 3.23), which may be summarised for a typical wood-boring insect as follows.

Most wood-boring insects are beetles, the adults of which lay eggs on the wood surface, in splits, in knots and so on. Upon hatching into larvae, these bore their way into the wood, creating a network of tunnels. The tunnels will be filled with excreted bore dust or frass. After a period of between one and five years the larva develops into a pupa and subsequently emerges from the wood, via a flight hole, as an adult insect. The features which will, in practice, aid in the identification of the insect type are:

- *Type and condition of timber*: insects of different species will favour different forms of wood – hardwoods or softwoods etc. Others will tend to infest only wood which is already affected by fungal attack.
- *Size and shape of flight holes*: the size and shape of the flight holes will inevitably be dictated by the size and shape of the insect.

Figure 3.23
Typical life cycle for wood-boring insects.

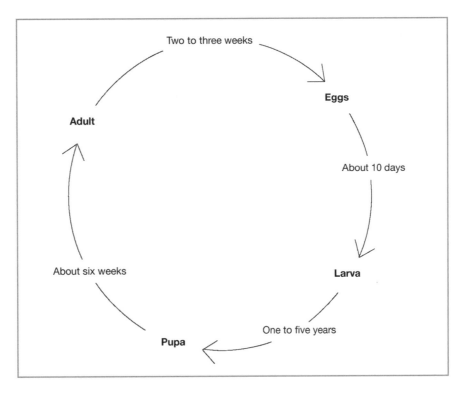

PART 2

- *Frass*: the shape, colour and texture of the bore dust or frass, often found beneath the wood after ejection via holes, give vital clues to identification.
- *Nature of bored tunnels*: the size and profile of the tunnels, which may be exposed by wear, again give clues to identification.

It is not appropriate to go into details regarding the wide variety of insect attacks that can afflict buildings within a text such as this. However BRE publications provide excellent, detailed appraisal of the various alternatives.

Reflective summary

- Timber is an inherently durable material and is resistant to most forms of degradation if maintained in a dry condition.
- The effects on timber arising from the attack of fungi and insects range from total physical deterioration to simple staining.
- There are several forms of fungus which can affect timber in buildings but which may not necessarily cause wood 'rot'.
- Fungal attack of timber in buildings can be split into two generic groups: wet rots and dry rot.
- The identifiable characteristics of different fungus types are linked to their life cycle and the stages of development which they undergo.
- There are two main categories of rot: brown and white.
- Within the category of brown rots is undoubtedly the most serious form of fungal attack: dry rot.
- The extent of damage caused by insect attack and the level of remedial action necessary are variable.
- The identifiable characteristics of insect species are linked with their life cycle.

Review task

Produce a matrix to compare the causes, features, identification methods of, and remedies for, dry rot and wet rot.

List the most common types of insect attack on timber, and explain how and when insect attack is likely.

3.8 | Dampness in walls

Introduction

- After studying this section you should have developed an understanding of the causes and effects of damp in buildings.
- You should also have developed an appreciation of how the type of damp can be identified and any remedial actions that may be required in order to solve the problems caused by the damp.
- You should also be able to identify features of new building work that have been incorporated to try to prevent damp being a problem in buildings.

Overview

The penetration of moisture to the interior of a building can lead to significant deterioration of the building structure and fabric. The comments made earlier with regard to timber decay are often associated with the failure of the building fabric to effectively exclude moisture. The existence of moisture or dampness in walls is one of the most common and potentially damaging building defects encountered. The effects of high levels of moisture upon timber, porous materials, metals and decorations can range from serious damage to minor cosmetic issues. As we strive to create more energy-efficient buildings with lower levels of ventilation, there is the risk of higher internal moisture levels and saturation of the air inside buildings. This may lead to problems if not managed appropriately. In general we can consider moisture in the walls of a building under four broad headings:

- Rising damp
- Penetrating damp
- Condensation
- Entrapped moisture.

PART 2

Rising damp

Many of the problems that are assumed to be the result of rising damp are actually more likely to be associated with hygroscopic salts giving the appearance of a damp wall.

The materials used to construct the walls of properties are generally porous; they contain large numbers of small voids or pores that have a tendency to attract moisture by capillary action. These materials are generally embedded in, or in contact with, the ground and in such situations the materials will encourage the migration of water from the ground, taking the form of rising damp. In addition, there is a natural action of osmosis that encourages the movement of water relative to the concentration of salts. Hence the movement of moisture up the wall depends on the natural porosity of the material and the presence of salts. It is also suggested that the electrical potential difference between the wall and the ground will have an effect.

The natural process of evaporation on the upper surfaces of the walls will have an effect upon the visible manifestation of the dampness. Materials that allow ready evaporation from the surface of the wall will encourage drying at the surface, and the dampness is likely to appear less significant and will diminish at a lower level. As the extent of evaporation is restricted by the use of impervious finishes etc., so the ability of the wall to shed moisture will be reduced and the dampness will rise to a higher level.

The extent to which a wall is affected by rising damp will vary from situation to situation. In traditional masonry walls, the limit of migration of moisture up the wall will be reached at around 1.2 m. However, this will be affected by several issues, including:

- The levels of moisture present in the ground
- The features that allow or restrict evaporation from the wall surfaces

- The extent of porosity of the material
- The chemical composition of the migrating water.

The water rising in the body of the wall will carry with it dissolved salts from the ground or from the masonry materials used to build the walls. These salts will be deposited in the pores close to the wall surface as evaporation of the moisture takes place and will have a series of effects. Firstly they may show physical salt presence in the form of surface efflorescence. Secondly they may block the pores of the material, thus inhibiting evaporation and encouraging the moisture to rise further up the wall. Thirdly, since they are hygroscopic in nature, they may encourage moisture from the internal environment (since all air will contain some water vapour) to be deposited at the wall surface, giving the appearance of dampness.

Causes of rising damp

The occurrence of rising damp is generally associated with older properties of traditional construction. In such properties it may well be the case that there was no effective damp-proof course installed in the original construction. However, there are many other causes of dampness, and it has been posited that fewer than 20 per cent of damp problems in walls are associated with true rising damp; of these only a small proportion are the result of the absence of an effective DPC.

Examples of other potential causes of rising damp include:

- Bypassing of the DPC caused by bridging internally by a porous floor screed
- Bypassing of the DPC externally by raised paths, planting borders etc.
- Bypassing of the DPC with external render coating
- Rain splashing on the external ground and passing above the DPC level
- Build-up of debris in a cavity allowing bypassing of the DPC
- Failure to link the DPC with the impervious membrane or DPM in adjacent solid floors.

The presence of rising damp will be indicated by a diminishing tidal pattern of moisture meter readings as we move up the wall. It is often the case that the readings will increase slightly at the upper extent as the effect of the hygroscopic salts takes effect, as noted earlier.

Penetrating damp

The occurrence of penetrating damp is highly dependent upon the levels of exposure of the building and it is often the case that moisture penetration occurs only on certain areas or elevations of the building. There are many reasons for the penetration of moisture in this manner, depending upon the materials and form of construction. In older, traditional buildings, the porous nature of the wall materials may be such that high levels of exposure simply force moisture through the wall. In more modern forms of construction, penetrating moisture tends to be associated with the failure of joints or jointing materials, allowing moisture to penetrate fissures, joints and cracks between areas of impervious cladding materials. The manifestation of such moisture problems tends to be associated with localised patterns of moisture on the internal wall surfaces and, unlike rising damp, can occur at all levels rather than being restricted to the lower regions of the wall.

Some typical examples of reasons for penetrating dampness include:

- Rain driving through exposed masonry walls that have insufficient thickness to resist the passage of water to the interior
- Problems associated with cavity trays in cavity walls
- Failure of joints in cladding systems
- Failure of rendered finishes
- Leakage of externally mounted rainwater goods
- Saturation of inappropriate insulation material in cavity walls.

Condensation

It is often the case that the internal environment of a building is warmer than the external environment. As such, the air within the building will hold greater levels of moisture vapour before reaching saturation. Any drop in the temperature of the air will result in the humidity of the air increasing; eventually this will result in the moisture contained in the air condensing. This is referred to as the *dew point*. When this occurs, moisture in the air will condense in the form of water droplets on colder surfaces. Hence the creation of surface condensation on the cold surfaces of a building interior may occur in certain conditions. This is often the case where localised areas of cold bridging are created by dense materials or by physical bridging across insulated cavities, for example.

The occurrence of condensation within the building fabric can also be an issue as the temperature drops below the dew point. This is referred to as *interstitial condensation* and is generally avoided by the use of vapour checks and ventilated voids. Condensation on the surface is sometimes mistaken for rising or penetrating damp, but is distinguished by being limited to the surface of the affected material and by having a distinctive pattern of moisture levels when noted with an electrical moisture meter (Figures 3.24 and 3.25).

Entrapped moisture

The construction process has traditionally relied on the use of 'wet trades', such as plastering and concreting, which introduce high levels of water into the building during construction. In the period following construction there will be a natural drying process, and this may take a considerable period of time before all of the construction moisture is removed from the building fabric. Hence the exis-

Figure 3.24
Patterns of moisture for sources of dampness.

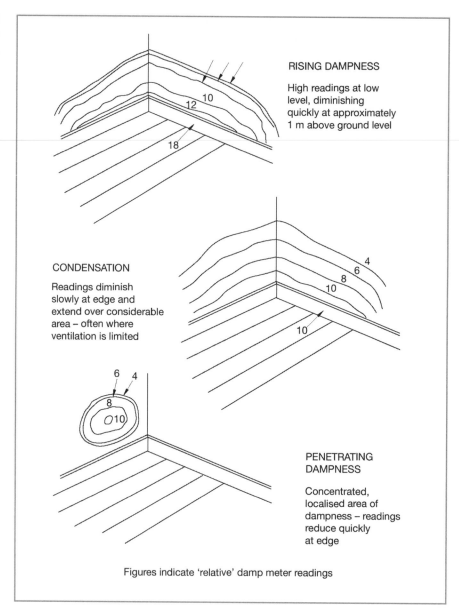

RISING DAMPNESS

High readings at low level, diminishing quickly at approximately 1 m above ground level

CONDENSATION

Readings diminish slowly at edge and extend over considerable area – often where ventilation is limited

PENETRATING DAMPNESS

Concentrated, localised area of dampness – readings reduce quickly at edge

Figures indicate 'relative' damp meter readings

tence of residual moisture in plastered surfaces, floor screeds and so on can be problematic in the early period of a building's life. The shift towards dry processes has alleviated this problem to a large extent, although some issues still need to be addressed in certain building elements.

Figure 3.25
Moisture meter readings for rising damp.

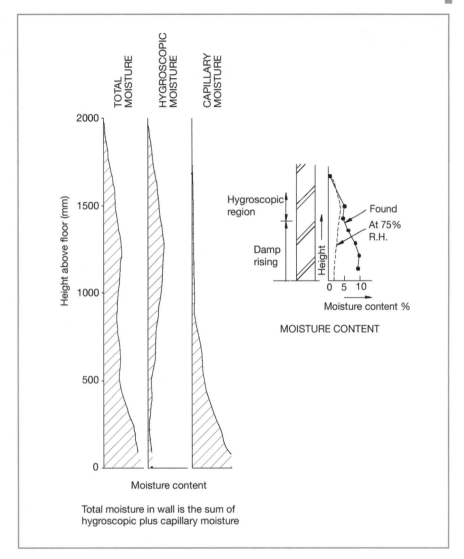

MOISTURE CONTENT

Moisture content

Total moisture in wall is the sum of hygroscopic plus capillary moisture

Reflective summary

- The penetration of moisture to the interior of a building can lead to significant deterioration of the building structure and fabric.
- The existence of moisture or dampness in walls is one of the most common and potentially damaging building defects encountered.
- In general we can consider moisture in the walls of a building under four broad headings:

  - Rising damp
  - Penetrating damp
  - Condensation
  - Entrapped moisture.

- The occurrence of rising damp is generally associated with older properties of traditional construction.
- In such properties it may well be the case that there was no effective damp-proof course installed in the original construction.
- The occurrence of penetrating damp is highly dependent upon the levels of exposure of the building.
- Condensation on surfaces is sometimes mistaken for rising or penetrating damp, but is distinguished by being limited to the surface of the affected material.
- The shift towards dry processes has alleviated the problem of entrapped moisture to a large extent.

Review task

List the main causes of damp in buildings and detail how the occurrence of damp can be avoided when constructing new buildings.

3.9 Flooding in buildings

Introduction

- After studying this section you should have an appreciation of the potential areas of damage to buildings arising from flooding.
- In addition you should understand the implications of flood risk to designers, owners and insurers of buildings, and you should recognise the key features that can mitigate risk.
- You should also be familiar with the broad design approaches now being adopted to reduce the risk of damage to properties that are at risk of flooding.
- In the event that a building suffers from a flooding incident you should appreciate the importance of early action and the sequence of remedial measures that could be taken.

Overview

In recent years the issue of flooding and its impact on urban areas has achieved a high profile. The increasing importance of climate change responses and the growing awareness of the need to consider sustainable approaches to dealing with climate change have prompted many studies into the risks and effects of flooding on buildings. Such is the extent of the concern that some residential areas that may be prone to the effects of flooding have seen major problems with property values and difficulties in procurement of appropriate insurance.

Flooding in context

Much guidance now exists relating to the mitigation of flood risk in design and, of most interest for this text, approaches to dealing with the effects of flooding after the event.

Whilst this text is not concerned with the design of new buildings as such, it is useful to refer to new building guidance to allow us to reflect on the issues arising from potential flooding risks during the life of a building. Four key aspects of designing buildings relative to flood risk are identified as follows:

- *Avoidance*: designing and locating buildings in such a way that they avoid the risk of being flooded.
- *Resistance*: designing buildings so that in the event of flooding taking place, water is prevented from entering the building and causing damage.
- *Resilience*: designing such that, in the event of water entering the building, the damaging effect upon the structure and fabric is minimised as far as possible.
- *Repairability*: designing in such a way that damaged elements can be readily and economically repaired or replaced following a flooding incident.

In the context of this text we shall limit consideration of flooding and buildings to the physical implications of water entry to buildings and the potential responses and remedial measures that can be employed. Whilst it is recognised that there is a vast array of guidance relating to policy and procedure in design and development relating to flooding, our primary focus here is dealing with the effects of flooding upon an existing building.

The causes of flooding can vary greatly and it is not the intention to consider these in any detail here since the physical effects on buildings are largely the same. However, the duration of flooding incidents can vary significantly and there is an implication arising from this, in that floods of greater duration will result in potentially greater levels of damage to buildings. The broad issue affecting the physical elements of a building are largely the same however. In order to consider these issues we must address three questions:

- How does flood water enter a building?
- What are the effects of flooding upon building structure and fabric?
- What are the remedial approaches to the effects of flooding?

PART 2

Modes of flood water entry

Water flow into a building behaves just like water flow through natural courses such as rivers in the sense that it will seek out the easiest pathway. As a result, water will penetrate buildings through open joints and voids quite readily. Depending upon the force and duration of flooding, water may penetrate through some or all of the following routes:

- Open joints around door and windows
- Gaps between different materials
- Gaps and voids around service penetrations
- Vents and airbricks
- Surcharging of drainage pipework and fittings
- Seepage through porous external wall/floor materials such as brick and concrete
- Flooding of cellars and sub-floor voids.

This is far from an exhaustive list but it illustrates the typical weak points in external enclosure that may lead to water entering the building.

Figure 3.26
Flood water entering a building.

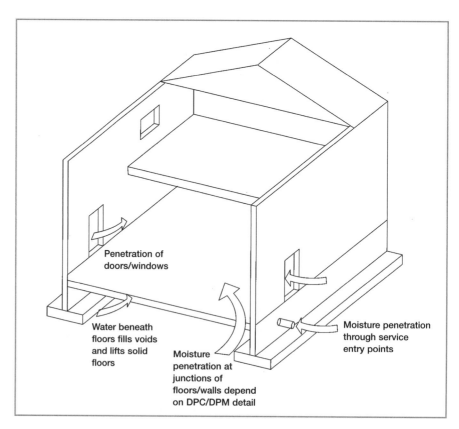

Penetration of doors/windows

Water beneath floors fills voids and lifts solid floors

Moisture penetration at junctions of floors/walls depend on DPC/DPM detail

Moisture penetration through service entry points

Effects of flood water entry

The extent to which flood water can damage a building depends upon the force, extent and duration of the flooding incident. Fast flowing water can cause significant impact damage affecting structural stability. However, this is thankfully quite rare in European climates where the majority of flooding damage arises from the wetting of the structures and fabric and the damage to building services. As a simple principle, the deeper the water and the longer the duration of wetting, the worse the effect will be.

Table 3.5 summarises the potential effects upon the structure, fabric and services.

Table 3.5 Potential affects of flooding.

Nature of flooding	Structure	Fabric	Services
Flooding below DPC level	Possible erosion of soil around foundations. Pressure beneath floor causes uplift	Corrosion of metal elements. Silt deposits. Saturation of sub-floor insulation. Residual mould growth	Silting of drainage system. Low-level fittings damaged
Flooding above DPC level (in addition)	Wetting of porous elements and potential corrosion of metal fixings	Damage to linings and finishes. Saturation of timbers and risk of future rotting. Saturation of insulation to walls. Damage to fixture and fittings. Surcharging of foul drainage system causing contamination	Damage to electrical distribution and fittings. Damage to floor-mounted gas appliances
Forceful/deep flooding (in addition)	Damage to walls and components from force of flowing water. Potential damage due to differential water levels against walls	Damage to higher level components and elements	Extensive damage to services installations

Remedial action for flooded buildings

The extent and nature of the remedial works required will vary from case to case and the technology of the specific building form will be key to establishing the essential remedial measures. One factor that is common to all flood affected buildings, however, is the need to effect drying of the structure in a controlled manner. This cannot commence until the water has subsided to an appropriate level. The timely intervention to address damage as soon as the water has subsided is critical in mitigating the extent of any damage that has been suffered.

A systematic approach is essential if the effects of flooding are to be dealt with effectively. Time is of the essence in dealing with flood damage and a deal of secondary damage can be avoided by timely and effective intervention in the period immediately following the flood.

Table 3.6 sets out the stages that will typically form part of the remediation process:

Table 3.6 Remediation process.

Stage	Actions	Time after flood incident
Initial damage assessment	Assess the extent of physical damage, possible structural damage and identify primary and potential secondary damage	Immediate
Prioritisation	Prioritise responses to elements of damage to reduce risk of further damage, structural failure and migration of water. Identify items that may be salvaged	Within a few hours
Clearance	Photograph and clear affected areas, remove salvageable items, dispose of damaged items. Remove any items that may inhibit remediation or process of drying	A few hours to a few days
Cleaning	Remove silt deposits, contaminated water and other detritus that has migrated into the interior, possibly hosing and jet washing of affected areas	A few days
Drying	Begin drying the property using ventilators, dehumidifiers and heaters from the top down	Several days to weeks
Remedial action	Address the remedial issues required to correct primary and secondary damage	Weeks to months

Initial drying

Returning the building to a dry state is the primary concern in the immediate period following a flooding incident. The degree to which drying out is achieved will dictate the further course of action and an acceptable level of 'dryness' must be defined. The general principle is that the building should be returned to its pre-flooding condition and that the moisture content of structure, fabric and components must be returned to the levels that existed prior to flooding. The indicators used to measure whether this has been achieved would be as follows:

- The materials or components within the building are in at least as good a condition as those that are unaffected by flood effects.
- The moisture content of building elements, components or materials is reduced to a level that will not support the growth of mould.
- The level of trapped moisture is reduced to a level that ensures that there is no risk of migration of water to areas that could result in mould growth or damage to areas that have been subject to repair or that were previously unaffected by the effects of flooding.

The process of drying must be carefully planned and monitored as there are risks associated with drying too slowly, such as fungal growth etc., or too quickly, such as material failure due to cracking/warping. Acceptable levels of moisture content in building elements must be pre-defined in order to reflect the pre-flood condition of the property and these will form the benchmark levels for acceptable dryness.

Primary and secondary damage

Having successfully returned the property to a satisfactory 'dry' state, the imperative is to deal with the physical damage caused by the flooding. The potential damage that may be suffered is categorised as :

- *Primary damage*: damage caused by the physical action of the flood process, including wetting of elements to the extent that they do not function as intended. Possible elements of primary damage might include:
  - Saturation of brickwork/blockwork
  - Silt deposits in cavities and voids
  - Saturation of timbers
  - Saturation of insulation materials
  - Lifting of screeds and floor slabs
  - Failure of electrical components and equipment
  - Surcharging of drains below ground
  - Damage to plasters, renders and finishes
  - Warping of timber boards, skirtings etc.
  - Debonding of tiles to walls and floors
  - Damage to decorative finishes
  - Damage to fixtures and fittings.

■ *Secondary damage*: damage resulting from the after-effects of flooding, such as wetting of adjacent elements and components, growth of mould and fungus due to failure to dry successfully etc. Possible elements of secondary damage might include:
  - Damage to areas unaffected by initial flooding due to migration of water
  - Residual high moisture levels in timbers etc.
  - Surface salt deposition resulting from drying
  - Mould and fungal growth.

Reflective summary

■ Flooding impacts on buildings and urban areas and there is growing awareness of potential flood risks.
■ Some residential areas have seen major problems with property insurance.
■ Four key aspects of designing buildings relative to flood risk are:
  Avoidance: Resistance: Resilience: Repairability.
■ The causes of flooding can vary greatly, the physical effects on buildings are largely the same.
■ Water flow into a building will seek out the easiest pathways, penetrating buildings through open joints and voids quite readily.
■ The extent to which flood water can damage a building depends upon the force, extent and duration of the flooding.
■ A systematic approach is essential if the effects of flooding are to be dealt with effectively.
■ The sequence of action following flooding will generally comprise:
  - Initial damage assessment
  - Prioritisation
  - Clearance
  - Cleaning
  - Drying.
■ Remedial action is necessary to address primary and secondary damage.
■ Primary damage is damage caused by the physical action of the flood process.
■ Secondary damage is damage resulting from the after-effects of flooding.

Review task

Identify the key features that might be problematic in the context of flood risk in your own home. Tabulate these features and summarise the response measures that might mitigate the level of flood damage.

Generate a 'timeline' that sets out the key actions and responses that need to be initiated in the period following a flood.

The technology of maintenance and refurbishment

4 Common refurbishment technologies

Aims

After studying this chapter you should have developed an understanding of:

The potential for extending the useful life of a building by undertaking refurbishment operations
The basis of the decision to refurbish
Some of the common operations involved in upgrading and refurbishing buildings
The reasons for the choice of various refurbishment techniques

This chapter includes the following sections:

4.1 Underpinning
4.2 Waterproofing of basements
4.3 Façade retention
4.4 Overcladding
4.5 Overroofing and reroofing
4.6 Upgrading and retrofitting of building services
4.7 Remedying dampness
4.8 Repairs to masonry
4.9 Treatment of timber defects

Info point

- BRE Digest 352 (1993): Underpinning
- BRE Digest 361 (1991): Why do buildings crack?
- Association of Specialist Underpinning Contractors (ASUC): Route to successful underpinning guide
- BRE Good Building Guide: Damp proofing existing building
- BS 8102: 2009 Code of Practice for Protection of Below Ground Structures (previously BS 8102: 1990 Code of Practice for Protection of Structures against Water from the Ground)
- British Wood Preserving and Damp Proofing Association (BWPDA): Code of practice for remedial water-proofing of structures below ground
- HSE guidance note 51: Façade retention (1992)
- Construction Industry Research and Information Services (CIRIA) Report 111: Façade Retention; Structural renovation of traditional buildings
- SC1 Publication 246: Over-roofing existing buildings using light steel (1998)
- BRE Report 185 (1991) Overroofing: especially for large panel system dwellings. Report
- Building Renovation Using Light Steel Framing (2008) SCI Yearbook
- European Liquid Roofing Association (ELRA) Guidance Note No. 1
- FRA Information Sheet No. 11: Maintenance and Refurbishment
- Building Services Research and Information Association (BSRIA) Technical Note 15/99: Retrofitting of heating and cooling systems

4.1 | Underpinning

Introduction

- After studying this section you should have refreshed your knowledge of why cracking occurs in buildings.
- You should also be able to identify which cause of cracking may lead to the need for underpinning to be used.
- You should be able to discuss and detail the different types of underpinning systems that are used, and outline what is good practice with regard to underpinning works.

Overview

The main reasons that buildings crack have been covered in the previous chapter, but can be summarised as:

- *Thermal effects*: different building materials expand and contract at different rates due to temperature.
- *Overloading*: when heavier loads are introduced into buildings than were expected when the initial design was undertaken.
- *Frost damage*: if there is any moisture in a wall, this will freeze in extreme temperatures, expand and cause cracks.
- *Vibration*: usually from roads.
- *Poor design* and/or construction.
- *Creep*: most buildings will deform slowly over time and cracks will occur.
- *Initial shrinkage cracks*: these occur while the building is drying out when newly built.

Poor design and construction account for approximately 60 per cent of all building defects and failures.

The cracks caused by the above tend not to be serious and would not require remedial works on a large scale to a building. However, when foundation movement occurs this may be to such an extent that extensive works are required.

Foundation movement can occur for the following reasons:

- *Settlement*: the load of the building compresses the ground underneath. This in itself may not be a problem if the amount of settlement is uniform across the whole building. However, if the settlement in the building is differential, i.e. one part of the building settles further than the other parts due to different ground conditions in that area, this can cause the building to 'drop' and cracks can be large. Remedial works may be required if the cracking is very unsightly or the settlement does not stop.
- *Heave*: this is upward movement of ground and usually occurs when the water content of the ground increases and the soil swells. This can occur due to very wet weather and poor drainage, flooding or the removal of trees from the surrounding area that would have dried out the soil naturally.

PART 3

■ *Subsidence*: this occurs when the ground under a building slips downward, causing the building to drop. This can occur because of:
  – Previous mine workings or old quarries
  – Excavation work that is undertaken near to existing foundations
  – Soil erosion caused by leaking drains
  – The soil drying out excessively due to long periods of very hot and dry weather, or an increase in the number of trees
  – Insufficiently compacted fill under foundations and floor slabs
  – The building collapsing into cavities in the ground where activities such as brine pumping have been carried out previously.

Tell tales can be used to monitor movement in buildings by measuring crack widths over time.

If foundation movement continues over a period of time and the problem cannot be resolved then underpinning may be required. This may be localised, and just the area where foundation movement has occurred will require underpinning. Alternatively, the entire building may require underpinning. This can be extremely expensive and may result in it being more economical to demolish the building and rebuild.

Underpinning systems

The methods used for underpinning domestic properties and multi-storey or large-span industrial and commercial buildings are largely the same, just on a different scale. The type of scheme used will depend on the scale of the remedial works required, the extent of damage, the ground conditions and the cause of the problem. The works will need to be designed by a structural engineer and undertaken by a competent and usually specialist contractor. Close supervision of the works will be required by the designer and building control officer in order to ensure the quality of the works. Health and safety can be a major problem when undertaking underpinning works, and well-detailed method statements need to be prepared before any work is undertaken.

Three very important procedures need to be undertaken when underpinning works are planned.

Drypack is a mixture of sand and cement, usually 1:3. Water is added to this mix to form a dry mixture that can be pushed into place without any of the sand and cement escaping. Usually a board is fixed to one side of the wall to cover the hole, and then the drypack is rammed into place to ensure there are no gaps. It is then left to cure.

1. Establish the cause of the subsidence through thorough investigation. This may be a desk study or may require some preliminary excavation.
2. Confirm the cause during the works. If it is different from what was originally believed to be the cause, then the scheme may need to change.
3. Confirm that the work being undertaken is actually addressing the cause of the problem, rather than just 'patching' up the problem.

Mass concrete underpinning

In this method, soil below the existing foundation is excavated out and a new foundation cast using concrete. The gap between the top of the new foundation and the bottom of the old foundation is then made good using drypack. Because

Figure 4.1
Mass concrete
underpinning.

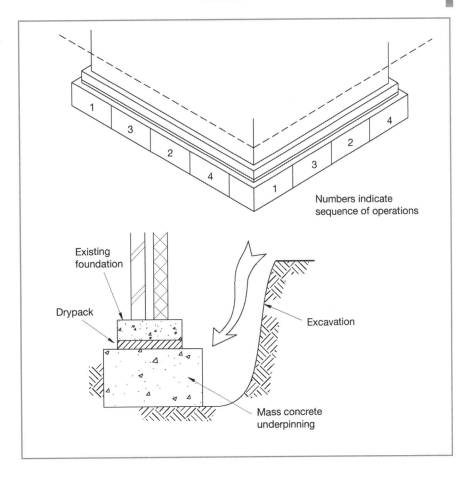

Numbers indicate
sequence of operations

Existing
foundation

Drypack

Excavation

Mass concrete
underpinning

the existing foundation will be unsupported whilst this work is being undertaken, the work is carried out in small sections. This is shown in Figure 4.1.

Sections 1 and 2 may be undertaken at the same time, and when complete sections 3 and 4 will be undertaken. The size of these sections will depend on the nature of the problem and the stability of the building. If the building is very unstable then these sections will tend to be quite small. If the base sizes are not sufficient in size then overstressing on the new base will occur.

Pier and beam underpinning

In this method a reinforced concrete beam is installed in the wall, either replacing the existing foundation or being placed just above it, as shown in Figure 4.2.

Once this is complete, excavation is undertaken below the existing foundation and voids filled with concrete. The gap between the existing foundation or beam and the new concrete is again made good using drypack. The advantage of this system is that less concreting and drypacking are required, but installing the beam can be complicated.

Figure 4.2
Pier and beam
underpinning.

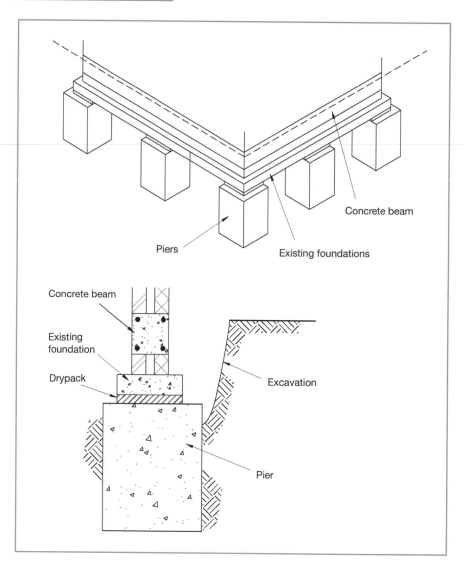

Concrete beam

Piers

Existing foundations

Concrete beam

Existing
foundation

Drypack

Excavation

Pier

It is good practice when
installing *in situ* concrete
piles to extend the pile by
400 mm and then break it
down to the correct level
when cured.

Pile underpinning

The type of piles used for pile underpinning can be either bored or driven. The
choice of system will depend on the ground conditions and access to the property.
In pile underpinning, piles are installed in one of the following ways:

- At each side of the wall (Figure 4.3(a))
- By one side of the wall (Figure 4.3(b))
- Inclined at both sides of the wall (Figure 4.3(c)).

The top of the pile should extend by a number of brick courses above the top
of the existing foundation. The tops of the piles are then connected using short

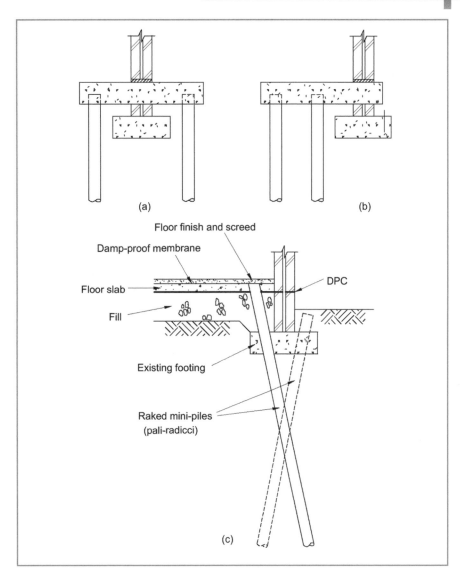

Figure 4.3
Pile underpinning.

reinforced concrete or steel beams that extend through the existing brickwork. The gap between the existing brickwork and the top of the beam is then made good using drypack. The advantage of piled systems is that no excavation of soil is required, and therefore there is less chance of a collapse while the work is being undertaken.

The advantage of using the cantilever system shown in Figure 4.3(b) is that no work is required within the building and therefore there will be little remedial work required inside the building. This will reduce costs and may also mean that the building can remain occupied while work is ongoing. The use of inclined piles, as shown in Figure 4.3(c), removes the need to break into existing brickwork in order to install a beam. Piles are drilled through the existing foundations and slab

PART 3

and take over the role of the existing foundation and slab support. The brickwork should remain intact and there will be little evidence of the work being undertaken.

Pile and beam underpinning

In this system piles are drilled next to the walls and beams inserted in both directions to join up the tops of the piles, as shown in Figure 4.4.

This is the most costly and complex system and is usually reserved for very serious cases. It will be impossible for a building to remain occupied while this work is being undertaken and a new concrete floor will be required.

Figure 4.5 shows details of a major underpinning scheme to a large building. Bored *in situ* piles have been cast in place either side of the existing outer wall. The tops of the piles have been connected using steel Universal Beam sections. Upstand beams have then been used to connect a new steel perimeter beam. This is supporting the existing wall. A new basement has then been excavated around the piles and the new foundation for the basement incorporated into the pile design.

Figure 4.4
Pile and beam underpinning.

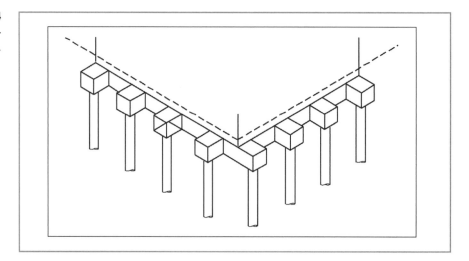

Good practice in underpinning

■ Any excavation should not be left open and should be concreted as quickly as possible. If the base is left open the soil could soften, and this will lead to settlement when the work is complete.
■ The underside of existing foundations needs to be well cleaned in order to ensure that the drypacked joint is sound and not contaminated with soil.
■ The quality of materials used needs to be as specified by the designer.
■ The works should not be rushed and method statements adhered to. Rushing could cause collapse.
■ Too much support must not be taken away at any one time.

Figure 4.5
Underpinning a large
building.

Reflective summary

- The main reason why underpinning is used is because of subsidence.
- Underpinning may be undertaken for the whole or part(s) of buildings.
- The systems used for underpinning are largely the same for domestic buildings and large-span, high-rise commercial and industrial buildings, only on a larger scale.
- Underpinning is very expensive and costs increase if there is damage to the interior of a building and/or building occupiers have to relocate while the work is being undertaken.
- The main systems use for underpinning are:
  - Mass concrete
  - Pier and beam
  - Piles
  - Pile and beams.
- Very competent contractors and designers are required for underpinning works.

Review task

Produce sketch details of:
  (a) Mass concrete underpinning
  (b) Pier and beam underpinning.

What are the three most common pile underpinning systems? List the advantages and disadvantages of these systems.

4.2 | Waterproofing of basements

Introduction

- After studying this section you should have developed an understanding of the reasons why the waterproofing of existing basements is undertaken and the benefits that can be derived from choosing this option.
- You should also be able to discuss and compare the different systems that are possible in order to achieve waterproofing.
- In addition you should be able to produce details of how reveals, partitions and ceilings need to be treated in order to achieve watertightness and be able to advise occupiers what precautions need to be taken when utilising waterproofed basement space.

Overview

The inclusion of basements in new domestic construction is very popular in northern continental Europe.

Constructing basements in the UK from scratch these days is not overly common in multi-storey commercial buildings, and virtually unknown in domestic properties. The reasons for this are that it is very expensive to construct basements and much more cost-effective to build an extra storey on top of a planned building to provide additional space. These days if basements are constructed they tend to be to provide car parking, or for a specific reason. For example, the British Library has seven storeys of basement, mainly to enable the internal conditions required for the storage of old and valuable books and manuscripts to be achieved. However, there are many older buildings including domestic properties that have basements, but they tend to have problems with damp. In order to make basements into usable space, damp proofing treatments need to be utilised. Because the basements are already constructed it is either not possible or would be very costly and difficult to waterproof them externally, and would require extensive excavation to the outside of the building. Therefore all of the works required for damp proofing existing basements need to be carried out internally. The cost of waterproofing needs to be carefully calculated and considered before the decision is made to carry out this activity. If the basement has been damaged due to flooding or a high water table then it needs to be decided whether waterproofing is in fact a feasible option, because damage in the future may be likely.

Other issues that need to be considered before the decision to waterproof is made include the ability of the basement to achieve the required thermal performance, ventilation options for the basement and the potential problems that could occur regarding condensation.

Features of damp proofing systems

Typical damp proofing systems include a number of elements, as shown in Figure 4.6.

Figure 4.6
Typical damp proofing
systems.

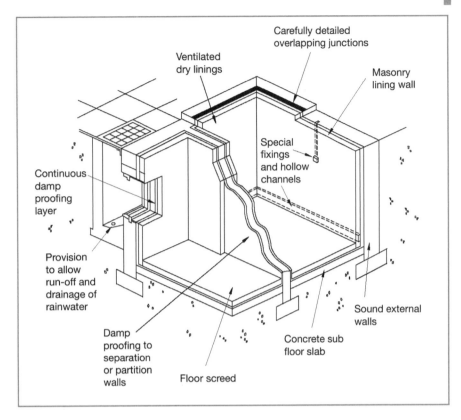

Carefully detailed
overlapping junctions

Ventilated
dry linings

Masonry
lining wall

Special
fixings
and hollow
channels

Continuous
damp
proofing
layer

Provision
to allow
run-off and
drainage of
rainwater

Damp
proofing to
separation
or partition
walls

Floor screed

Concrete sub
floor slab

Sound external
walls

Concrete sub-floors and sound, solid external walls are required to provide a good solid base to apply damp proofing to. An effective and continuous damp proofing layer with carefully detailed junctions between wall and floor damp proofing is required, and this must tie in with existing DPCs. Overlaps of 100–150 mm are recommended. If the risk of moisture ingress is low then the use of ventilated dry linings can provide adequate protection. Damp proofing needs to continue into basement partition walls and the basement treated as a monolithic waterproof structure as opposed to a series of parts. If this occurs there will always be a weak point and water will find a way through the damp proofing. Floor screed or a concrete slab is required to protect the damp proofing from damage through use, and suitable wall linings are required to protect wall damp proofing from damage and keep it in position. Special fixings for services are required in order to prevent penetration of the damp proofing due to drilling of the wall, and provision for ground water run off must be built into the design of the waterproofing system.

Before any damp proofing work is specified, the cause of existing damp needs to be identified. Damp may be caused because existing damp proofing is missing, damaged or unsuitable. Alternatively, it may be caused because the water table is high. This may have occurred naturally, it may have risen due to excessive rainfall, or it may be that existing drainage is damaged or is inadequate. Condensation may be the cause of the damp problem due to poor ventilation or heating levels, and leaks from existing pipework over time, or due to a burst pipe, may have caused or be causing the damp. If the cause of the damp problem is condensation or leaking pipes, damp proofing is not the correct course of action to remedy the situation. If condensation is the main issue then ventilation and heating of the basement need to be improved via the installation of a HVAC system. Alternatively, altering the building structure to allow for natural ventilation could be an option. Existing pavement lights could be removed and opening windows incorporated into the wall to the basement. Figure 4.7 shows how this could be achieved.

If this approach to the introduction of ventilation into the basement is to be taken, the structural implications need to be carefully considered, especially if the basement walls are loadbearing. This is also an issue when considering partition walls within the basement. It may seem that the easiest option is to remove the partition to facilitate damp proofing, but this may not be feasible because of limitations of the structural support capability of the external walls.

If plumbing leaks are the causes of the damp, then pipes need to be repaired or replaced and insulated to avoid further leaks. The basement needs to be dried either naturally or by using dehumidifying machines before redecoration is undertaken.

It is good practice to remove all existing timber, such as lintels and door linings, from the basement and replace them with non-timber alternatives. This is because the damp timber may not dry sufficiently and/or may distort when the damp proofing is completed; or if sealed behind damp proofing the timber might rot.

Figure 4.7
Incorporation of opening windows into existing basements.

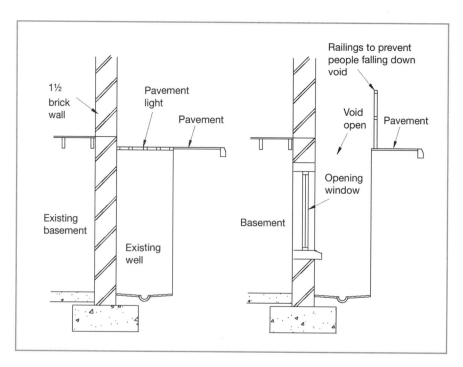

Basement damp proofing systems

DPC – Damp-proof course.
DPM – Damp-proof
membrane.

■ *Drained cavities*: in this system all floor and wall finishes need to be removed, including plaster and screed. A row of engineering bricks needs to be laid with open joints at intervals, leaving a 50 mm cavity between them and the existing wall. A DPC needs to be laid on top of the engineering bricks and a new block-work wall built that is tied with stainless steel wall ties to the existing wall. A sump and a self-draining tile layer are then installed and covered with a sheet of DPM. A new screed layer is then applied over this system. This system is effective if installed correctly, but does reduce the space of the basement due to the installation of the wall and the cavity. See Figure 4.8(a).

Figure 4.8
Alternative basement damp
proofing systems.

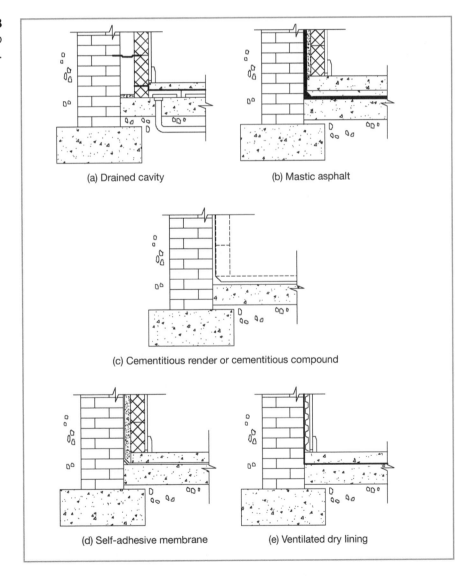

(a) Drained cavity

(b) Mastic asphalt

(c) Cementitious render or cementitious compound

(d) Self-adhesive membrane

(e) Ventilated dry lining

PART 3

- *Mastic asphalt*: in this system all floor and wall finishes are removed, including plaster and screed. The existing floor needs to be scabbled to give a rough surface and existing brickwork joints raked out to provide a key for the asphalt. A two-coat asphalt fillet is then applied to the floor/wall joint followed by three coats of asphalt to the floor and two coats to the walls. A protective screed is then applied to the floor, followed by a reinforced concrete slab. If this has a power float finish then no further finish will be required other than a decorative cover, such as carpet. A new internal wall is then built backed onto the asphalt to the walls, which is backfilled with sand and cement as the wall is constructed. This system also has implications for the size of the room in all dimensions. It is a costly system, but ultimately if constructed correctly is very durable. See Figure 4.8(b).

- *Cementitious render*: in this system, all wall coverings must be removed and surfaces scabbled/raked out and cleaned so that there is no trace of any previous finishing. Cement corner fillets are provided at wall/floor and wall/wall junctions. Three layers of proprietary mix are then applied either by trowel or by spraying, allowing recommended times between the layers for curing. The walls can then be skimmed with plaster to give a finish for decorations. With this system the floors are usually replaced separately before the walls are treated, and the inclusion of DPMs and DPCs needs to be carefully detailed. This system is very durable and reduces the restriction of space that the previously discussed systems incur, but the cementitious layer can be easily damaged by drilling or hammering fixings into the walls. See Figure 4.8(c).

> Cementitious renders are usually sand and cement mixes with waterproofing agents added.

- *Self-adhesive membranes*: in this system existing brickwork needs to be cleaned down and flush pointed or skim rendered to provide an even surface. Floor coverings need to be removed down to sub-floor level and concrete cleaned and thoroughly dried. Then the wall/floor and wall/wall fillets need to be formed, followed by the application of the membrane to the walls and floors, which basically just sticks to the existing surfaces. Sufficient overlaps at joints need to be allowed. The membrane is then protected by a layer of blockwork backfilled with sand and cement to the walls, and a floor screed of minimum thickness 50 mm to the floor. See Figure 4.8(d) for details. This system is a durable option and is good for the fixing of services etc., but there will be a reduction in space due to the additional wall.

- *Liquid applied membrane*: the preparation for this system involves removing all flooring down to sub-floor level and then cleaning the slab thoroughly and allowing to dry. Wall brickwork needs to be flush pointed and then cleaned down. The application and detail of this system are the same as for the self-adhesive system, except that the material is wet as opposed to dry. Suitable materials for the liquid membrane include bitumen emulsions and epoxy resins. This system is a durable option and is good for the fixing of services etc., but there will be a reduction in space due to the additional wall.

- *Ventilated dry linings*: preparation for this system includes brushing off all loose wall material and flush pointing of brickwork up to 150 mm above the slab, followed by the removal of all floor finishes down to sub-floor level. The floor and walls then need to be thoroughly cleaned. A DPM is then laid on the floor and allowed to turn up the wall by at least 150 mm. A new floor screed of at least 50 mm needs to be laid to protect the DPM.

Dimpled plastic sheets are then fixed to the wall using fixings recommended by the manufacturer. Gaps of 10 mm are needed at both the bottom and top of the sheets to allow for ventilation. The sheets are then plastered or plasterboard is attached, ensuring that ventilation gaps are maintained. See Figure 4.8(e). This system does not significantly reduce room sizes but is only suitable if the basement has a slight damp problem. The sheets are also susceptible to damage.

Whichever system is used, it is advisable to finish walls using moisture-permeable paint that will allow water in any wet applied coatings to dry out completely. Special care needs to be taken by occupants to ensure that damp proofing layers are not damaged, and heating and ventilation levels need to be carefully monitored to avoid problems with condensation. It is very important that building occupiers are advised against installing any extra services without gaining advice from specialists. For example, if a cementitious damp proofing system has been used, this is a monolithic covering that protects all external and partition walls. If this material is damaged due to someone cutting holes or chases, then water collecting behind the wall will use this as a path into the building. In the worst case scenario, the whole of the damp proofing system will require replacement. More common problems with waterproofed basements include occupants fixing shelves or pictures to the walls using screws, which damage the waterproofing layer and cause localised damp problems.

Dealing with partition walls

When partitions are non-loadbearing it is advisable to remove them, complete the waterproofing, rescreed the floor and rebuild the partition.

When partitions are loadbearing and there is no requirement to remove them, the usual approach is to continue the damp proofing specified for the external walls onto the partition walls. Figure 4.9 illustrates this scenario

Figure 4.9
Waterproofing of partitions in existing damp basements.

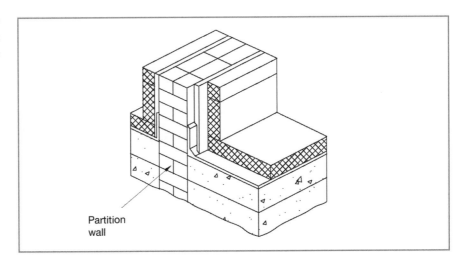

Partition wall

Dealing with basement ceilings

Where the basement wall extends above ground level, the ground level DPC needs to be checked for effectiveness and position. If the existing DPC is deemed to be faulty or ineffective it must be replaced as part of the damp proofing works. Basement damp proofing systems must be continued up to the ceiling level to provide a level surface for finishings. Ventilation needs to be built into new ceilings, and existing joist systems and fixings may need to be isolated from the main structure by using joist hangers. Figure 4.10 illustrates these recommendations.

Dealing with thresholds and reveals at external openings

Damp-proof courses should be introduced at all thresholds and around all external reveals, such as door and window openings. All materials used for windows and doors should be suitable to withstand the effects of moisture. It is good practice to isolate these elements from the main walls.

Fixing of services

If the waterproofing system chosen incorporates a lining wall then services can be installed in the normal way. If ventilated dry linings are used, then moisture-resistant fittings need to be used and a waterproof seal used at the outlet, as illustrated in Figure 4.11(a). When cementitious damp proofing systems are used, services are run in recesses in the wall that have been cut sufficiently deep and wide enough to allow for the placement of the cementitious material and the installation itself. This is shown in Figure 4.11(b).

Figure 4.10
Ceilings in basements being waterproofed.

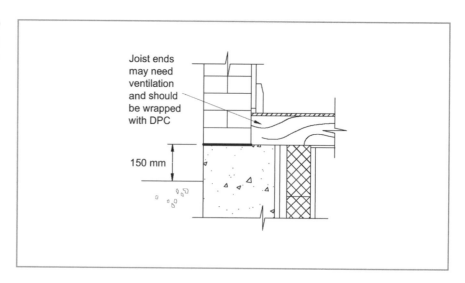

Figure 4.11
Fixing of services in
waterproofed basements.

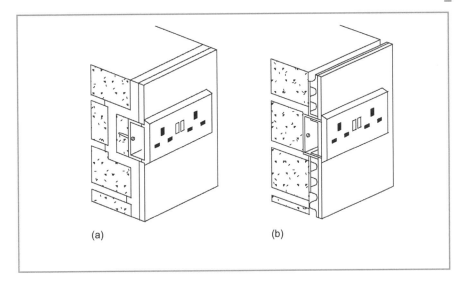

(a)   (b)

Reflective summary

- There are many older buildings, including domestic properties, that have basements, but they tend to have problems with damp.
- Existing basements are already constructed and therefore it is either not possible or would be very costly and difficult to waterproof them externally, and would require extensive excavation to the outside of the building.
- All of the works required for damp proofing existing basements need to be carried out internally.
- Issues that need to be considered before the decision to waterproof is made include the ability of the basement to achieve the required thermal performance, ventilation options for the basement and the potential problems that could occur regarding condensation.
- Before any damp proofing work is specified, the cause of existing damp needs to be identified.
- The use of waterproofing systems is not suitable if the main cause of damp in basements is leaking pipes or condensation.
- Basement damp-proofing systems include drained cavities, mastic asphalt, cementitious render, self-adhesive membranes, liquid applied membranes and ventilated dry linings.
- Special attention needs to be given to the details at reveals, thresholds, partitions, ceilings and points where services are fixed to ensure that these do not become weak points that will encourage localised water seepage in waterproofed basements.

Review task

Produce a comparative study matrix for different systems that can be used to waterproof basements, and compare each system against the following criteria:
- Durability
- Potential for use where high water table exists
- Skill levels required for installation
- Space penalty
- Health and safety issues.

PART 3

4.3 | Façade retention

Introduction

- After studying this section you should have developed an understanding of the reasons why façade retention schemes occur.
- You should have gained a knowledge of the different options for façade retention and be able to determine the best option for different circumstances.
- You should understand why differential settlement occurs and be able to explain the reasons why movement should be allowed for when designing fixings for façade retention projects.
- You should also be able to explain why problems with access can occur during façade retention work.

Overview

The façade retention systems discussed here are only used when any of the external walls are the main loadbearing elements of the building.

Façade retention is the term used when some or all of the existing external walls of a building are retained during a refurbishment while the internal structures are removed, leaving a hollow shell or even a single external wall. The benefits of this are linked to the ability to create a new internal layout free from the restrictions of a loadbearing cellular construction form. This is popular as a design solution where there are planning restrictions affecting a building. The ability to replace the building core while retaining the external façade permits the creation of a modern building environment suited to users' needs with a traditional external appearance.

Systems of façade retention

Several systems are available for providing the required temporary support during façade retention schemes. The choice of system depends on a range of factors, including:

- Accessibility of the site
- Degree to which pavements or roads may be obstructed
- Structural form and condition of the existing façade
- Amount of building structure to be retained
- Scale and proportions of the retained sections.

Initially, an assessment of the existing façade is required in order to determine the most appropriate system to use. The following will need to be investigated before any choice can be made:

- Are there any defects internally, externally or within the façade? Some defects may be obvious, such as cracking, but there are others which may be hidden.

- Is the façade loadbearing or is the main support for the building a structural frame?
- Which materials have been used to construct the façade?
- Have any remedial works been undertaken previously, including remedial works to foundations?
- How plumb is the façade? Modern buildings are generally designed as perfectly square and if the façade bows or tilts forwards or backwards then any modern building work behind the façade will have to be designed accordingly. Also, the stability of the façade during this work could create further problems.

Structural issues

It is very common when carrying out façade retention schemes to find that the façade also needs underpinning because of the inadequacy of the existing foundations.

The vast majority of new multi-storey buildings constructed these days use a structural frame as the main support for the building. The frame transmits all of the dead, live and weather loadings from the external façade, roof and floors to the foundations. If a structural frame is used then façade retention techniques as such are not applicable. Three sides of the building cladding can be removed without affecting the stability of the façade that is remaining (Figure 4.12). However, in older buildings the external walls are commonly the loadbearing part of the structure and they are tied together by the floors. Loads are transmitted to

Figure 4.12
Process of façade retention.

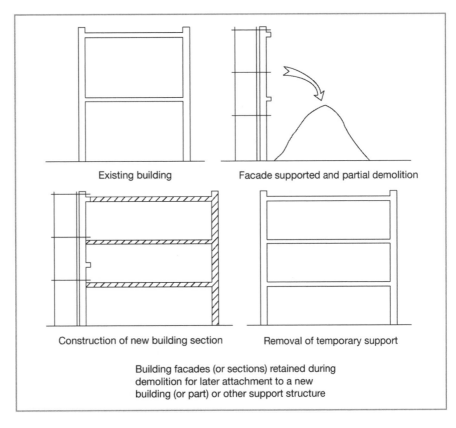

Existing building

Facade supported and partial demolition

Construction of new building section

Removal of temporary support

Building facades (or sections) retained during
demolition for later attachment to a new
building (or part) or other support structure

PART 3

Figure 4.13
Underpinning to existing
foundations.

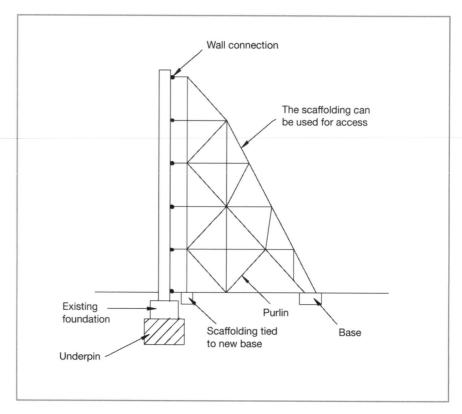

the foundations from the floors and roof via the external walls. Therefore if you remove the floors, the walls are liable to collapse, and thus they need to be supported during this operation.

The new frame will then take the structural loads from the building, independent of the façade.

Another problem that is common in older buildings is that the original foundations are not adequate and do not comply with current Building Regulations. Any façade that is to be retained is therefore likely to need underpinning (Figure 4.13), or at least the foundations extending.

In the majority of façade retention schemes the existing internal structure is going to be replaced. The replacement structure will generally take the form of a structural frame constructed using *in situ* concrete or structural steel. The new frame must be designed to support all loadings plus the existing façade(s) and transmit them to the foundations.

Façade retention options

The options available for façade retention support systems are essentially of two types – internal support and external support. The most common forms may be summarised as follows:

Figure 4.14
Internal and external raking
shores.

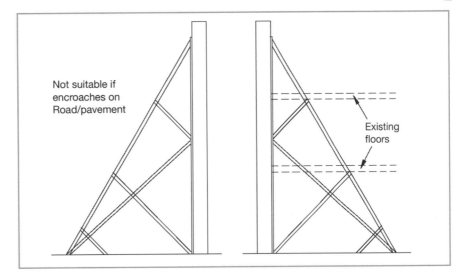

- *Raking shores*: these may be fixed internally or externally and comprise a series of triangulated braced supports providing lateral support to the retained wall (Figure 4.14).

  Figure 4.15 shows external raking shores being used to support a loadbearing brickwork façade.
- *Horizontal bracing*: this may take the form of beams or trusses running horizontally along the elevation braced by tower sections. This is a very common form of retention system. A foundation is formed and columns connected to this. A temporary steel frame is then constructed and the façade fixed to it during construction work. Figure 4.16 shows an example of this.

  This may also take the form of continuous scaffolding running horizontally and vertically to create a 'birdcage' support system.

Figure 4.15
External raking shores.

Figure 4.16
Horizontal bracing
(*far right*).

Figure 4.17
Horizontal cross-bracing
options.

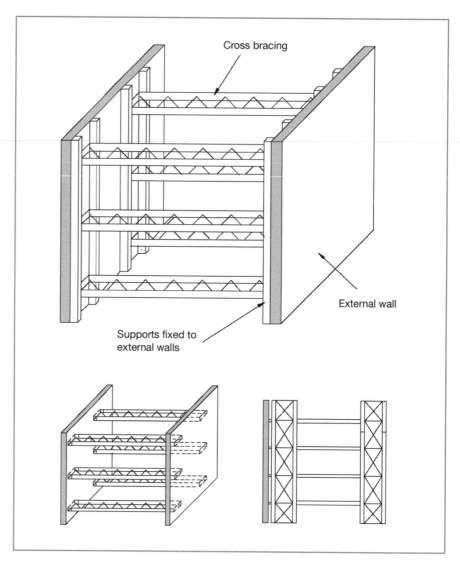

- *Internal cross-bracing*: this takes the from of a series of internal struts or braces, often in conjunction with one of the foregoing systems to provide a rigid support system where the entire exterior of the building may be retained. One of the limitations of this approach is the degree to which access for working is afforded internally. The struts (Figure 4.17) can be:
  - Bedded into the existing walls
  - Connected to walers attached to the walls
  - Connected to towers that are independent of the walls.
- *Façades strutted across the building*: horizontal forces are transferred to the flank walls. In this case the restraint given to the walls by the floors is replaced by the bracing systems. The struts are fixed to the two walls to restrain them securely together (Figure 4.18).

Figure 4.18
Mutual support of existing walls.

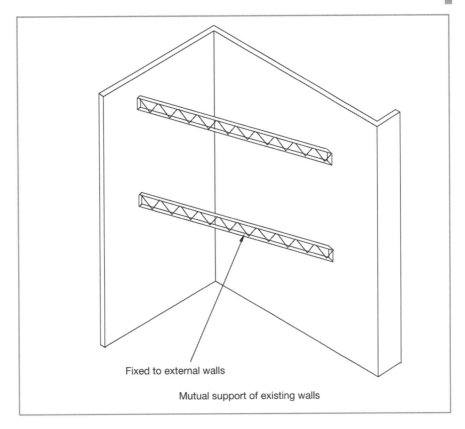

Fixed to external walls

Mutual support of existing walls

- *Internal façade systems that employ the use of the new steel frame*: utilising the new steel frame as the façade retention system (Figure 4.19) reduces the cost of the works considerably. The new steel frame columns are installed after holes in the existing floors have been broken out, and then the beams are fixed to the back of the façade. The floors can then be demolished. This is only suitable if the new floor levels do not clash with the existing floor levels.
- *Proprietary systems*: there are companies that specialise in the manufacture and installation of proprietary systems. One of these is RMD Ltd, who use Slimshor components (Figure 4.20). The advantages of these systems are that they are easy and quick to install and use standard components, ensuring a shorter lead-in period for delivery.

Differential settlement

The term *differential settlement* means different elements of a building or adjacent buildings moving at different times or rates.

Differential settlement can be a problem in façade retention schemes. It occurs when construction work that has been undertaken at different times settles at different rates. For example, it may take a building up to two years to settle when newly built, but after this initial settlement there should be no further movement unless ground conditions change, such as where clay caps dry out and heave

Figure 4.19
Utilisation of new steel
frame for retention of the
façade.

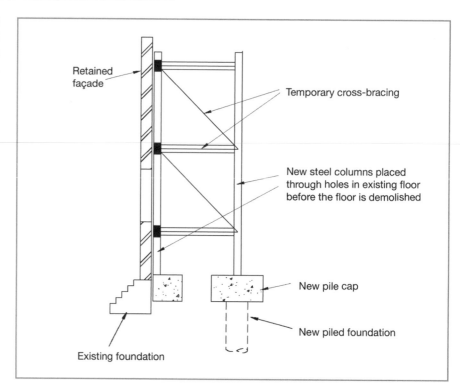

Retained façade

Temporary cross-bracing

New steel columns placed through holes in existing floor before the floor is demolished

New pile cap

New piled foundation

Existing foundation

occurs, or if drains crack and the resulting seepage washes away the soil support-ing the foundations.

In façade retention schemes, the façade will tend to be well established and all initial settlement will have occurred. The new build work is then completed and must undergo its initial settlement phase. Thus the existing façade will remain static but the new build will move. If the connections from the existing to the new work do not allow for some movement between the two structures, potentially serious structural damage will occur. This is not a major problem when using temporary fixings, but the permanent fixings need to incorporate features that prevent differential movement. Slippage between the existing wall and any encased steelwork should be allowed for by the use of slip joints between the two elements (Figure 4.21). These joints will allow for movement of the new work and prevent the new work from causing structural damage to the existing struc-ture. Incidentally, slip joints are required when new buildings are constructed adjacent to existing buildings, regardless of whether the contract is a façade reten-tion scheme.

Fixing types

Temporary fixings are required during construction work only: they do not form part of the finished building. Temporary fixings usually consist of a collar inserted

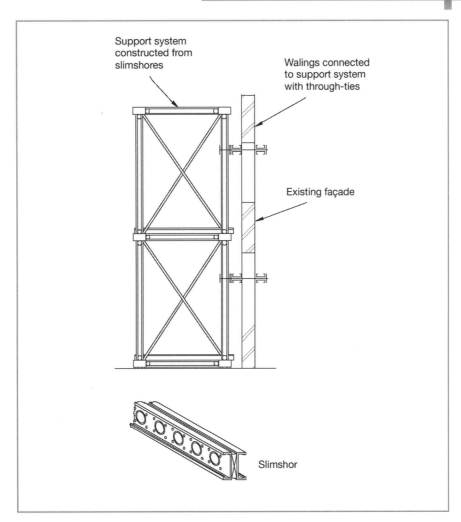

Figure 4.20
RMD proprietary system.

Support system constructed from slimshores

Walings connected to support system with through-ties

Existing façade

Slimshor

on either side of the wall connected to the truss system. These fixings are then tightened and wedges used to take up any unevenness of the wall.

Permanent fixings can be achieved using a steel plate and rod fixed to the beam by an angle section rather like a sliding anchor. Irregularities in the wall can be accommodated by increasing the number of fixings. Again, as with sliding anchors, angles with slotted holes can be used to allow for differential movement. The bolt to the external wall is disguised using a brick slip, which is the same colour as the external brickwork or stone. Figure 4.22 shows some temporary and permanent fixings.

Alternatively, fixings can be made using resin anchors. A hole is drilled and filled with resin (Figure 4.23). The bolt is then placed and when the resin is dry the connection can take place. The disadvantage of this type of fixing is that they are rigid and will not allow for differential settlement. They are therefore commonly used for temporary fixings only.

Figure 4.21
Methods of forming slip
joints.

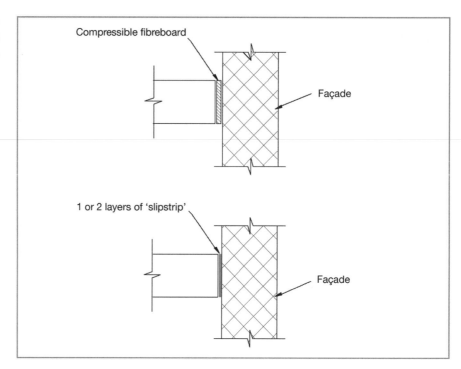

Access issues

When undertaking façade retention schemes there are usually problems with access. Access can be divided into two sections:

■ Access for construction work
■ Avoiding encroachment on public highways or footpaths.

Access for construction work can be severely restricted when using an internal façade retention system. Initially there can be problems installing the system before major demolition can occur. Holes will need to be broken through floors and excavation through ground or basement floors will be necessary in order to form foundations to support the retention system. Once the system is fixed in place and the demolition has occurred, it may be difficult to use machines to clear the debris because of the façade retention supports. Some hand removal of demolished material may be necessary. During the construction of the new structure, care has to be taken so as not to damage the support system, and new work may have to be built around the supports, and the 'gaps' filled in later. Removal of the support system can also be problematical, especially if large trusses or girders have been used. It may be impossible to lift the sections out once the façade is fully supported by the new structure, or they may have to be cut or burned out in sections. If it is at all possible, an external façade retention system is therefore preferable. Figure 4.24 shows the advantage of using an external system. Access to undertake the new build work is completely unrestricted.

Figure 4.22
Temporary and permanent fixings.

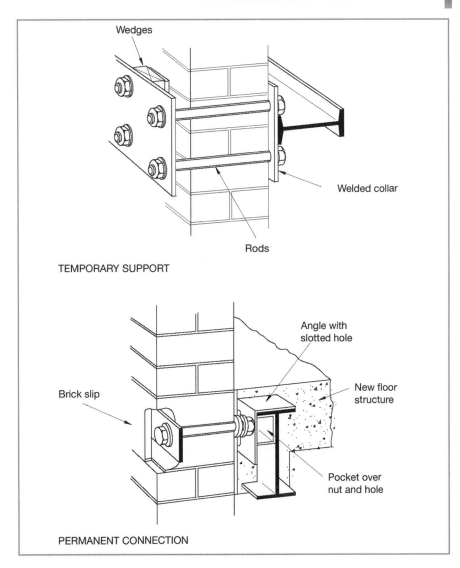

Wedges

Welded collar

Rods

TEMPORARY SUPPORT

Angle with slotted hole

New floor structure

Brick slip

Pocket over nut and hole

PERMANENT CONNECTION

However, when using an external support system, this will certainly encroach on any pavement and maybe even into a road. This will create problems for the public and could pose a health and safety issue. These problems could be reduced by supporting the façade on a series of frames that allow for walking underneath. The frames could be formed from steel sections or scaffolding and could then be cased in plywood to improve the appearance. Fire protection can be applied to the steel before casing in the plywood casing (Figure 4.25).

The holes in the existing façade would be cut out, with the steel sitting on a sand and cement base in the slot. The steel beam would then be placed and drypack used to make up the gap between the steel beam and the existing façade material. This forms a very strong support for the retained façade onto the supporting steel. Once the drypack is fully cured, the façade will be retained by a series of these portals.

Figure 4.23
Example of fixings using resin anchors.

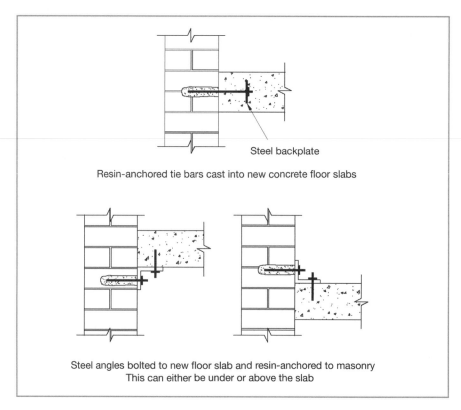

Steel backplate

Resin-anchored tie bars cast into new concrete floor slabs

Steel angles bolted to new floor slab and resin-anchored to masonry
This can either be under or above the slab

Figure 4.24
An external support system.

It is important to remember that this will not support the façade on its own and an external support system (Figure 4.26) will be required to be built over the top of the portal. Figure 4.27 shows an example of how access to a footpath has been preserved for the duration of the construction work.

Figure 4.25
Steel frame used to support the façade.

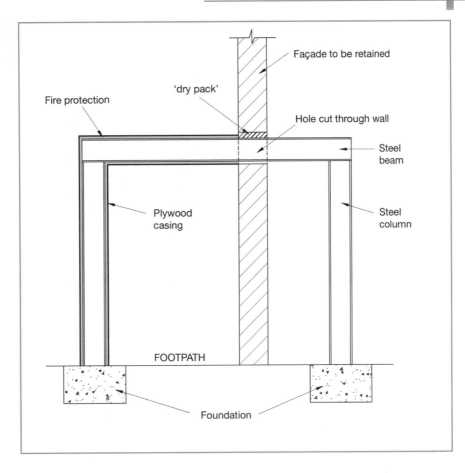

Figure 4.26
External temporary support system.

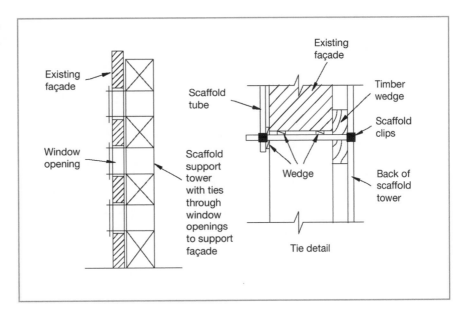

Figure 4.27
Preserving access to a
footpath.

 Comparative study: Façade retention

Longitudinal 'birdcage' structure

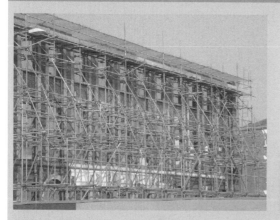

Figure 4.28 Façade retention using scaffold birdcage.

Here we see the front elevation of a terraced property supported using a framework of scaffolding poles to create a 'birdcage' structure. The structural form of the retention system is essentially that of longitudinal bracing with the scaffold structure providing the framework for support. Note the heavy concrete ballast units that are used to add mass to the structure to resist movement.

Figure 4.29

Figure 4.29 shows the framework structure with scaffold poles passing through existing window openings to retain the structure to the support framework. The framework is externally positioned to allow access to the rear of the elevation for rebuilding.

The bracing of the existing wall and securing to the support structure are effected by longitudinal members connected to

sections of scaffold tube that pass through the window openings (Figure 4.30). Note the use of timber planks to either side of the openings. These provide additional bracing and rigidity to the brickwork, which is in poor condition and could not be relied on in terms of structural integrity.

The bracing tubes (Figure 4.31) pass through holes that are created in the elevation in positions where there is not an existing opening such as a window. The assembly of the supporting scaffold is clearly visible here.

Figure 4.30   **Figure 4.31**

Proprietary system

Figure 4.32   **Figure 4.33**   **Figure 4.34**

The system illustrated in Figure 4.32 is based on longitudinal bracing with supporting towers formed from a proprietary steel system.

   Figures 4.33, 4.34, 4.35 and 4.36 show the connection of the support bracing to the existing structure. The 'Slimshor' system provides longitudinal bracing to front and rear faces of the existing wall (timber blocks are used as packing/contact protection against the existing stone). The braces are then connected using threaded studs that

Figure 4.35   **Figure 4.36**

clamp the façade to provide support. Note the new steel frame that has been erected behind the façade and extending along the frontage; this will be connected to the existing façade to provide permanent support.

PART 3

- *Façade retention* is the term used when some or all of the existing external walls of a building are retained during a refurbishment while the internal structures are removed, leaving a hollow shell or even a single external wall.
- Several systems are available for providing the required temporary support during façade retention schemes.
- Systems can be classified as internal or external.
- Initially an assessment of the existing façade is required in order to determine the most appropriate system to use.
- When the external walls are the main loadbearing element of a building, if the floors are removed the walls are liable to collapse; thus they need to be supported during this operation.
- Differential settlement can be a problem in façade retention schemes.
- When undertaking façade retention schemes there are usually problems with access.

Review task

What factors will affect the choice of façade retention system?

What needs to be considered during the initial assessment of a façade if retention of the façade is proposed?

4.4 | Overcladding

Introduction

- After studying this section you should have developed an understanding of the reasons that building owners may choose to overclad existing buildings.
- You should be able to explain the different classifications of overcladding systems, and list the main features of the different systems.
- You should also be able to describe the potential problems that can occur in overcladding, list the component parts of overcladding and produce typical details in sketch format.

Overview

There are many existing buildings that have a structurally sound frame, but in which the external façade has deteriorated to the extent that it has become dangerous, i.e. parts of the cladding are falling off. Alternatively, the façade may be deemed to be old-fashioned in appearance, and in this circumstance a building owner may decide to give the building a 'facelift' for aesthetic purposes only, or because of the extent of the deterioration of the external cladding. However, there are other reasons why the external fabric of the building may be deemed to be unfit for purpose. A major problem that building owners may have in the future is that they may wish to undertake a major refurbishment of the building, and they may be satisfied with the building aesthetically in the main, and have no health and safety problems. Nevertheless, when plans are submitted for approval to Building Control, they will have to show compliance with Part L of the Building Regulations, and they may not be able to achieve this without substantial works to the fabric of the building. Compliance may only be achievable if the building is overclad. Overcladding is a completely new skin applied to the whole or part of the external walls to upgrade the performance and/or appearance of the original building. Whilst most cladding materials are suitable, care is needed in their choice and application.

Reasons for choosing overcladding

The major reasons why building owners may choose to overclad buildings are:

Spalling can be caused when water has made contact with the steel reinforcement, which then rusts and expands, resulting in the concrete blowing off.

- *Inadequate weathertightness of the external envelope*: water may be seeping into the building through cladding panels themselves, through joints in external cladding or through leaking window and door units. These are repairable in the majority of cases and therefore the cost of overcladding may not be justified. However, joint repairs can deteriorate and may then have to be repaired again, so a long-term view may be that overcladding is in fact the most cost-effective option.
- *Deterioration of concrete and external finishes*: cracking and spalling concrete, together with falling surface finishes, can be dangerous to passers-by and look unsightly. Overcladding can eliminate these problems and give the building a completely new look. Figure 4.37 shows a case in which the aggregate in rendered infill panels has fallen off, which is both dangerous and unsightly.

Figure 4.37
Loss of aggregate rendering.

PART 3

Figure 4.38
Position of insulation in overcladding systems.

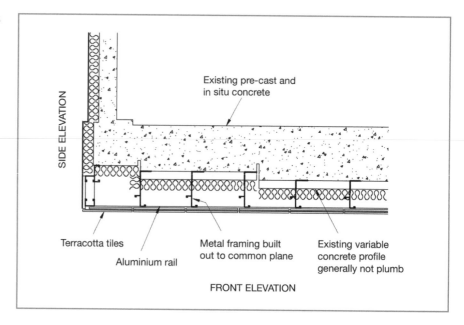

SIDE ELEVATION

Existing pre-cast and in situ concrete

Terracotta tiles

Aluminium rail

Metal framing built out to common plane

Existing variable concrete profile generally not plumb

FRONT ELEVATION

- *Improving thermal insulation*: many buildings were designed before the sustainable construction debate started, which calls for greater energy efficiency of buildings, and well before the stringent requirements of Part L of the Building Regulations came into force. The external walls of buildings now have to possess far better U values in order to comply with these regulations and in most existing buildings this will not be the case. Therefore in order to upgrade buildings the thermal performance of cladding needs to be improved. Insulation can be incorporated into overcladding systems. However, with enhanced insulation in poorly heated buildings, condensation can increase due to the lowering of circulation rates as a result of the improved sealing of the building.
- Figure 4.38 shows how insulation has been incorporated into an overcladding system. The existing building was clad in precast concrete cladding panels that have deteriorated badly and are uneven due to spalling of the fascia. The existing fascia is also stepped to form part of the original design feature of the façade. A metal framing has been used to build out the line of the façade to a common plane and new terracotta tiles attached to the framework. The insulation has been fixed directly to the existing façade.
- *Improving appearance*: many large buildings designed in the 1960s and 1970s are viewed as ugly eyesores. The appearance of a building can be totally changed by overcladding. Overcladding can be designed using the majority of materials and the building appearance changed dramatically. Very futuristic or very traditional façades can be produced.

Figures 4.39 and 4.40 show façades before and after overcladding for two different buildings. The photographs are black and white and may not highlight the huge change to the appearance of the building, but in the second set the

Figure 4.39
An example of overcladding to change a building's appearance.

Figure 4.40
Another example of a change of appearance through overcladding.

new cladding is orange, compared with the initial grey render finish of the building.

■ *Reducing noise levels*: improving the thermal insulation of a building can also improve the sound insulation. However, the inclusion of an insulating layer is only one of the elements that supports the improvement of sound insulation. Sound transmission reduces if layered construction is used, and overcladding systems tend to be composite, and therefore layered, which will increase sound insulation and reduce noise penetration into the building.

Approaches to overcladding

The broad principles of overcladding are the same as those for the initial cladding of buildings. Several options exist, that reflect the generic categories of traditional cladding. It is entirely possible, for example, to utilise brick cladding as a method of overcladding. However, it is in the nature of building refurbishment and modernisation that there is generally an intention to improve the physical performance of the external envelope, principally in terms of thermal performance, and to bring about a more modern appearance. Also, the implications of additional load upon the building structure must be carefully considered. For these reasons the technologies most commonly used for overcladding are those associated with true cladding or rainscreen cladding forms.

PART 3

As with new buildings, there are three generic forms of true cladding systems. These are:

- *Sealed or impervious systems*: in these systems the external envelope is made to be a continuous, impervious enclosure with panels supported by a lightweight framework and with the joints between panels sealed using gaskets or some other form of sealant. The high-tech, mirrored facades of many buildings of the 1980s result from this form of cladding.
- *Drained systems*: these systems adopt the approach of impervious panels over-lapped at their junctions to create a largely impervious envelope but with the capacity for minor water penetration at the joints to be drained through a cavity or void and directed to the exterior. The examples shown in the Case study photographs (Figures 4.44 to 4.47) illustrate this approach. Limited airflow is facilitated through the void and the cladding is compartmentalised into a series of vertical sections.
- *Rainscreen cladding (pressure equalised)*: these systems utilise air-pressure in the void to manage the effects of wind action on the building façade. They gener-ally have baffled, open joints, without seals or gaskets, aimed at minimising moisture penetration due to driving rain. Positive pressure is promoted within the cavity to enhance moisture evaporation and there will be an impervious layer on the outer face of the existing building façade to resist any potential moisture transfer.

Like true cladding and rainscreen cladding, the systems are generally panelised and are supported by a skeleton framework (as with curtain walling) attached to the structural frame of the building. Some, less sophisticated systems may be secured directly to the outer face of the existing buildings but these are quite rare.

Strength and stability of overcladding systems

Any applied wind loadings will affect the overcladding system itself, and if the overcladding system is porous then some of the wind can seep through to affect the structure itself. When assessing wind loads, the overcladding can be consid-ered to be in one of two categories:

- Installations with a void between the overcladding and the building: usually panels are fixed to battens or a grid of cladding rails attached to the building. This is shown in the front elevation of Figure 4.38.
- Installations with no void between the overcladding and the building: usually insulating panels bonded to the building surface, with an impermeable outer skin. This is shown in the side elevation of Figure 4.38 and in the window detail shown in Figure 4.41.

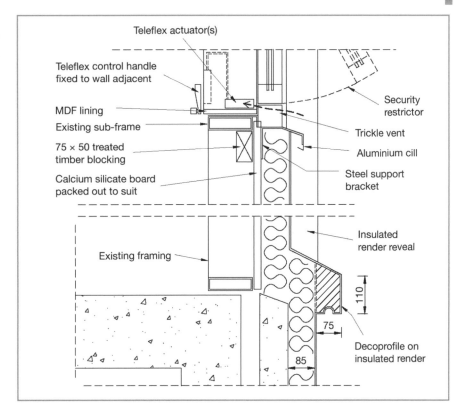

Figure 4.41
Overcladding window
detail.

Overcladding systems with voids

There are three principal loading cases for overcladding systems with a void:

- *Overcladding impermeable, voids vented to a known location*: in this case the pressure in the void will equilibrate to the pressures at the void vent. Usually the void is vented externally and the pressure in the void will equilibrate to the external wind pressure. Cavity barriers can be located to divide the void into manageable sections.

- *Overcladding moderately permeable, void large*: in this case the pressure in the void will equilibrate to the average of the external pressure over the area of the overcladding, and the load on the cladding will be the difference between the external applied wind pressure and the internal void pressure. Therefore the smaller the area between cavity barriers, the smaller the cladding loads will be. Cladding loads are minimised if cavity barriers are placed in positions where changes in external pressure occur such as at the corner of buildings.

- *Overcladding very permeable, void small*: in this case the pressure in the void cannot equilibrate and there will be a significant flow through the void from areas of higher external pressure to areas of lower pressure. This sets up a gradient of pressure in the void, where the void pressure is always closer to the local external pressure than in the previous systems, thus relieving the load on the cladding.

Cavity barriers are a very important part of overcladding systems.

PART 3

Overcladding systems with no void

These systems are usually impermeable to enable them to resist rain penetration. In most cases the degree of permeability of the building surface is indeterminate and it is safest to assume that the overcladding must transmit the full wind loads through any adhesive bond or mechanical fixing

Possible problems in overcladding systems

- *Fatigue:* continual flexing of panels and fluctuating wind conditions can lead to fatigue and subsequent cracking. Fluctuations in external surface temperature can also lead to fatigue of material at the interface in metal-skinned sandwich or laminated panels. These fluctuations can lead to local delamination, which can affect appearance and durability.
- *Impact damage*: in practice, accidental impact damage to high-rise buildings occurs infrequently, but damage can occur through the use of access cradles. At lower levels there may be damage due to vandalism and graffiti. The stability and integrity of the whole wall are normally assessed by two types of impact test: 'hard' body and 'soft' body. The soft body test assesses the ability of the cladding to resist a heavy blow from a large impactor: this could be from an access cradle or fire ladders, and the test only fails if the cladding becomes dislodged. A hard body (steel ball) impact of 10 N m is also required for the whole surface of the overcladding, with the same criterion that nothing should fall off. BS 8200 gives impact energies of 500 N m up to 1.5 m above access, and 350 N m above that height. There is no requirement for soft body impacts above 1.5 m.
- *Overloading of the foundations*: if the existing cladding is left in place, the additional loading from the new cladding must not overload the foundations.
- *Weathertightness*: rain leakage through joints in overcladding is the main reason for failure. Rain water will absorb into the cladding until it is saturated and run-off begins. Some water will bounce off the surface and be absorbed back into the atmosphere, and water can be driven upwards in windy conditions. Overcladding a relatively absorbent building with an unabsorbent skin will change the run-off characteristics of a building, and a building that did not need a canopy over the entrance may now need one. The use of a rain screen allows for rain penetration to drain down the cavity into catchment trays without reaching the structure.
- *Fire*: if non-combustible materials are used for insulation or overcladding, the risk of fire affecting the cladding is low except when the actual structure is a risk. However, there may be a risk of lightning damage, as overcladding may change the existing arrangements. This will need to be considered in the design of the new system.

The required life of the cladding system should be the same as the whole building – 60 years is the normal expectation. However, it is difficult to predict how the

Figure 4.42
Components of over-
cladding systems.

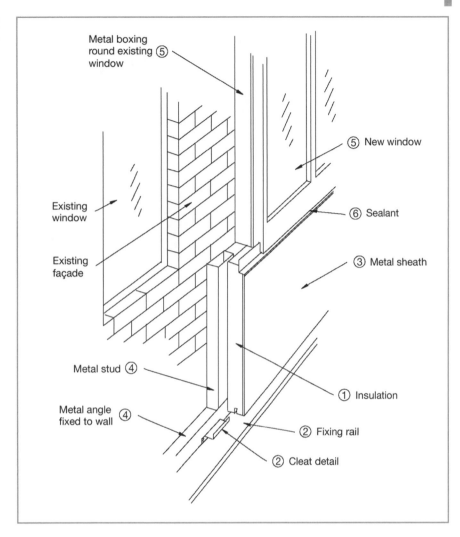

Metal boxing
round existing ⑤
window

⑤ New window

Existing
window

⑥ Sealant

Existing
façade

③ Metal sheath

Metal stud ④

① Insulation

Metal angle
fixed to wall ④

② Fixing rail

② Cleat detail

PART 3

cladding will behave over time because of the following:

■ Some materials are prone to damage due to location. For instance, it is unsuitable to use aluminium near the sea as salt water can seriously affect the material.
■ Levels and type of local industrial pollution can change over time, causing
degradation of the material.
■ Extreme temperatures can affect the material and, although climate changes
are generally slow, there can be 'freak' bouts of extreme weather.
■ Colour changes can occur, and some materials, such as GRP, are more prone to
this after prolonged exposure to ultraviolet light.
■ Maintenance levels may change – although some materials may claim to be
self-cleaning, very few are totally self-cleaning under the action of rain water.

Figure 4.43
Overcladding in practice.

Figure 4.43
Overcladding in practice.

Therefore if expenditure on maintenance falls, the overcladding may need replacing earlier.

The component parts of overcladding

Full overcladding usually consists of six basic components:

GRP stands for glass reinforced plastics.

- Thermal insulation
- Fixing for insulation
- Outer skin or weatherproof layer – this can be lath and render, thin plastics-based render, battens and sidings of wood, metal or uPVC, tile hanging, brick-work, sheet metals, GRP panels etc.
- The suspension and fixing system for the outer skin
- Windows
- Weatherproof joints.

A typical system is shown in Figures 4.42 and 4.43.

Case study

Overcladding

The Case study photos show an overcladding project on a 1960s education building. The existing exterior was clad in a mixture of pre-cast concrete panels and brickwork, with steel casement windows. The condition of the windows was poor and the thermal performance of the building was far below current requirements. Hence, it was decided to overclad the exterior with an aluminium rainscreen system, incorporating new windows and increased levels of thermal insulation.

Figure 4.44 shows the original concrete cladding panels and the dilapidated windows. The vertical aluminium cladding rails have been fixed to the outer face of the wall in preparation for the fixing of the overcladding.

Here in **Figure 4.45** we see cladding installers fixing the supporting cladding rails to the existing façade, utilising a cradle to assist with safe access.

Figure 4.46 shows clearly the insulation secured to the face of the existing cladding, set between the cladding rails for the new system. The aluminium rainscreen system is secured to the new rails to provide a robust weather shield.

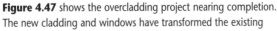

Figure 4.47 shows the overcladding project nearing completion. The new cladding and windows have transformed the existing building both aesthetically and in terms of functional performance.

PART 3

4.5 Overroofing and reroofing

Introduction

- After studying this section you should have developed an understanding of what the differences are between overroofing and reroofing.
- You should be able to outline the reasons why overroofing is chosen as an option and the different systems that are possible.
- You should also have a knowledge of the reasons why roofs fail and the options for repair that are possible.

Overview

The term *overroofing* relates to the operation of providing a new roof structure and covering over the top of an existing roof. One of the most common approaches is the creation of a new pitched roof assembly over the top of an existing flat roof.

Failure of flat roofs is a common problem. The reasons for this could be faulty design, construction, maintenance or materials, or a combination of these.

In such instances the provision of the new roof is an alternative to the recovering or total replacement of the original, possibly failing, flat roof. This activity is termed *reroofing*.

Overroofing

There are several reasons for the consideration of overroofing in addition to simply repairing a defective existing roof. Typically the reasons for considering overroofing may include:

- Rectification of long-term flat roof failure that has had a great deal of money spent on it during its life through corrective maintenance.
- Improving weathertightness. Roofs must be capable of withstanding wind, rain and snow, and probably in most instances it will be water penetration that is the problem. Water can be airborne, in the form of rain, or run-off water. Weathertighness will be improved if any joints are kept clear from the flow of run-off water and/or above the run-off water level. This is shown in Figure 4.48.

   Another factor that is important in achieving improved weathertightness is the quality of the surface fixings. Figure 4.49 shows the screw fixings sealed with mastic. It is important that these fixings are installed perpendicular; if they are skew then the weatherproof washers will not be compressed evenly and will become a weak point.
- Aesthetics: pitched roofs are often considered more attractive than flat roofs. Figure 4.50 shows how the addition of a pitched roof can change the appearance of the building. The advantage of the pitched roof being positioned below the parapet is that the wind loadings on the building will not be seriously affected.

PART 3

Figure 4.48
Joints in roof covering above the potential run-off water levels.

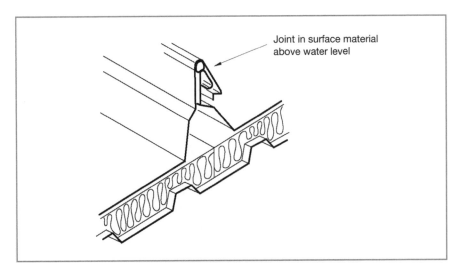

Joint in surface material above water level

Figure 4.49
Screw fixing to overroofing panels.

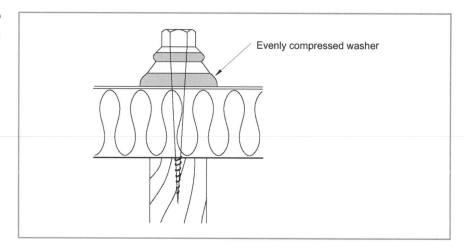

Evenly compressed washer

Figure 4.50
Pitched roofs on flat roofs.

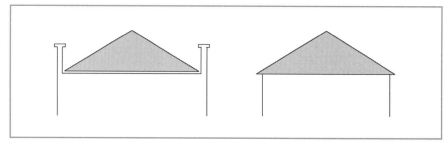

- Minimising disruption in building operation: overroofing can take place without disturbing the building interior.
- Improvements in the thermal insulation properties of the roof. It is possible to improve the thermal insulation of a roof by applying insulation internally to the underside of the roof slab. However, the insulation becomes very warm because it is heated by the warmth of the building, but the concrete slab will be very cold. This could lead to condensation either on it or within it. It is therefore more advantageous to apply insulation on the top of the concrete deck, and if installed correctly this will avoid cold bridging. External insulation can be protected by an overroof structure or by a cladding system. Figure 4.51 shows how this can be achieved in sandwich and composite forms.
- Creation of extra usable space, which could include habitable space or space for services in a void created by pitched trusses to support a new roof. Figure 4.52 shows a good example of how this has been achieved.
- A reduction in long-term maintenance costs.

In most cases a combination of these factors drives the decision to select overroofing as an alternative to simply repairing an existing roof, although it must be stressed that overroofing is sometimes used when the existing roof is not defective, but simply for the reasons noted above. Perhaps the most common approach to overroofing is the utilisation of a lightweight steel structure clad with profiled metal cladding.

Figure 4.51
Addition of thermal insulation to existing roofs.

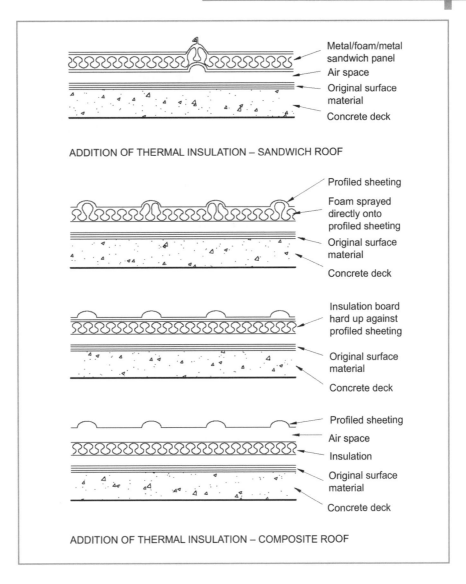

Metal/foam/metal sandwich panel
Air space
Original surface material
Concrete deck

ADDITION OF THERMAL INSULATION – SANDWICH ROOF

Profiled sheeting
Foam sprayed directly onto profiled sheeting
Original surface material
Concrete deck

Insulation board hard up against profiled sheeting
Original surface material
Concrete deck

Profiled sheeting
Air space
Insulation
Original surface material
Concrete deck

ADDITION OF THERMAL INSULATION – COMPOSITE ROOF

The benefits of this approach include:

- The ability to cater for long spans due to the high strength-to-weight ratio of steel
- The ability to provide a lightweight roof without imposing significant extra load on the building
- Flexibility of form
- Robustness and durability of the construction form
- Ability to accommodate a variety of claddings and finishes.

The more flexible the form, the more possible aesthetic options there are.

The decision to adopt overroofing as a technical design solution must take into account a number of functional factors. The intended purpose of the finished roof

Figure 4.52
Adding an extra storey by
overroofing.

Figure 4.52
Adding an extra storey by
overroofing.

assembly and the space within it will dictate some of the design decisions that are made. In general it must be accepted that the roof design should be such that it provides maximum strength with minimum weight. In addition, the ability of the existing building structure to cope with any additional loading must be taken into account, along with the mechanisms for providing structural connections between the new roof and the existing structure. Another significant issue is the ability to cope with the required modifications to rainwater disposal systems. This arises due to the fact that flat roofs tend to drain via internal rainwater outlets. This is obviously not possible with a pitched roof solution and alternative approaches must be adopted.

Structural forms of overroofing

There are several commonly adopted structural forms for the creation of over-roofing assemblies. The selection process depends on the form of the existing building and the nature of the space required within the finished roof void. In general they can be considered to fall within three broad classifications:

- Rafter and purlin roofs
- Trussed roofs
- Structural frames.

Rafters and purlins

Perhaps the simplest approach to overroofing is the adoption of a traditional rafter and purlin roof structure. Where there are suitable cross-walls or supports for locating the purlins they can be positioned to span between, with common rafters connected to the purlins to provide the pitched roof structure.

Trusses

Overroofing using prefabricated trusses allows the roof to be supported from the exterior walls or structural frame of the building.

This form of overroofing has two discrete approaches. Firstly there is the possibility of forming a roof structure using multiple closely spaced trusses, suitably braced, to support the roof deck or covering.

Secondly there is the potential to provide more robust trusses at wider spacings which support purlins that, in turn, support the roof deck or covering.

Structural frames

In some situations it is important that the area of usable space generated as a result of the process of overroofing is in the form of a genuine extra storey on the building. This can be facilitated by the formation of an overroofed area using a moment-resisting frame. Here the use of robust steel members to provide a frame

Figure 4.53

Comparison of the different structural forms commonly used for overroofing.

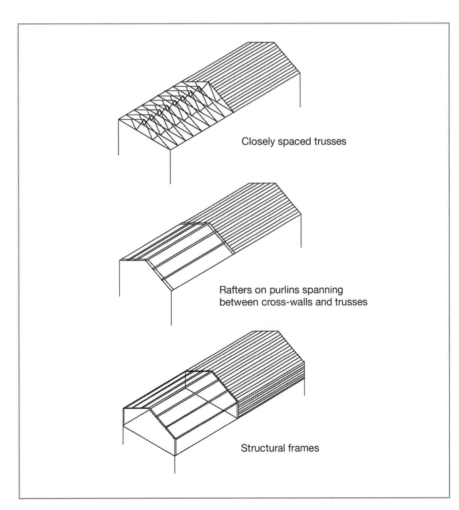

Closely spaced trusses

Rafters on purlins spanning between cross-walls and trusses

Structural frames

PART 3

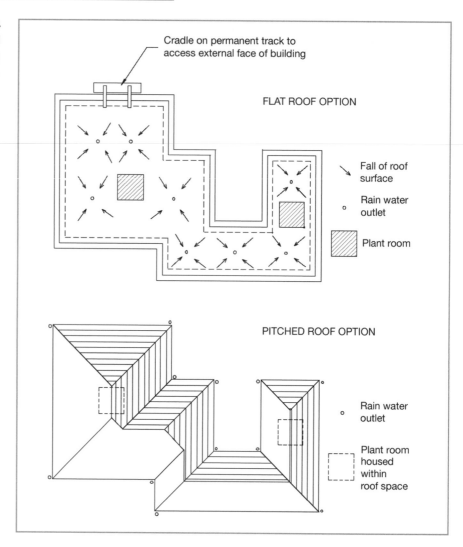

that does not need internal bracing is appropriate, as it ensures usable storey
height space. Figure 4.53 shows the differences between the three most common
systems.

It is possible to overroof using pitched or flat roofing systems for any building,
including very complex shapes. Figure 4.54 shows flat and pitch roof options for
a complex building plan.

Reroofing

Reroofing may be required for a number of reasons and it may be that repairs to
the existing roof are feasible, and overroofing is not necessary. The most common
reasons for roof failure and the methods used to repair roofs are as follows:

Figure 4.55
Correct positioning of
insulation to avoid cold
bridging.

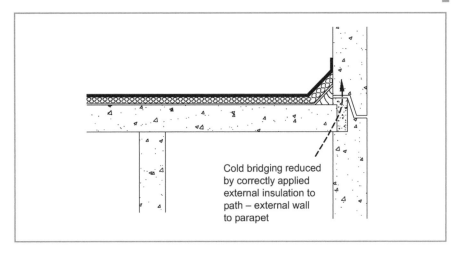

Cold bridging reduced
by correctly applied
external insulation to
path – external wall
to parapet

- The roof is leaking through the existing membrane or deck in localised areas. Spot repairs to the membrane may be undertaken if the affected area is not extensive, or the whole roof may need to be stripped and a new membrane applied. Additionally, there could be leakage through rooflights, flashings, copings or damp courses, and these can be replaced.
- The roof is ponding because the original design did not include adequate falls, or there has been some deflection of the structure or failure of an existing deck or insulation. This may require removal of the original coverings and making good to the areas that are sagging, after which the membrane can be reinstated. Alternatively, a liquid waterproofing system may be used.
- There is condensation because of insufficient insulation, the building is very humid, there is no ventilation between the roof deck and ceiling in a cold roof construction or there is a lack of a vapour control layer. The existing coverings could be removed and new insulation applied before recovering. The insulation should be applied correctly, as shown in Figure 4.55 in order to avoid cold bridging.
- There are cracks or splits in the roof covering caused by impact or overloading damage and/or because of deterioration of the covering due to age. Again this could be rectified by undertaking localised repairs, or alternatively by applying a liquid waterproofing system.
- Blistering of the waterproofing membrane due to the effects of temperature. This is likely to affect large areas of the roof and therefore the application of a liquid waterproofing system may be the best option, or the roof could be completely stripped and a new membrane applied.

Liquid waterproofing systems

The main advantages of liquid waterproofing systems are the lack of disruption, as the existing substrate is usually left in place, the homogeneous nature of a

PART 3

Asphalt, which is a commonly used roofing material, will soften considerably in hot weather.

seamless coating, the ability to cope with irregular surfaces, and the ease of installation. This is basically an overcoating system, and is suitable for use if:

- The substrate is clean and dry
- The existing deck and any insulation is dry and structurally sound
- The existing membrane will bond with the new material
- The existing deck will take the additional loadings
- The height of upstands and details will allow for the additional thickness of the waterproofing material.

If any of these conditions are not satisfied it may be necessary to strip the existing roof and relay the finishes.

Liquid waterproofing systems can be laid over bituminous and non-bituminous membranes and coatings, profiled sheets, concrete, slates and tiles, and can be used on flat or pitched roofs.

Reflective summary

- The term *overroofing* relates to the operation of providing a new roof structure and covering over the top of an existing roof.
- *Reroofing* refers to the repair or replacement of an existing roof.
- There are several reasons for the consideration of overroofing in addition to simply repairing a defective existing roof.
- Overroofing is sometimes used when the existing roof is not defective in order to improve the properties of the roof.
- Structural forms of overroofing include rafter and purlin roofs, trussed roofs and structural frames.
- Reroofing may be required for a number of reasons and it may be that repairs to the existing roof are feasible and overroofing is not necessary.
- The main advantages of liquid waterproofing systems are the lack of disruption as the existing substrate is usually left in place, the homogeneous nature of a seamless coating, the ability to cope with irregular surfaces, and the ease of installation.

Review task

For what reasons might overroofing be preferred to reroofing?

What are the advantages of using liquid waterproofing systems to recover a roof, and in what circumstances can such a system not be used?

4.6 | Upgrading and retrofitting of building services

Introduction

- After studying this section you should have developed an understanding of the term *retrofitting of services*.
- You should also be able to explain why it may be necessary to retrofit services installations a number of times during the lifetime of most buildings.
- You should be able to list the principles of retrofitting and discuss the scale of intervention that may be required in given scenarios.
- You should also be able to demonstrate an understanding of how legislation could influence the choice of design for a retrofit services installation.

Overview

As the complexity of business information technology advances, so the ability of commercial buildings to cope with its incorporation must be managed more proactively. Modern buildings rely on sophisticated environmental control services and the ability to cope with IT infrastructure in order to be commercially competitive. For these and other reasons it is unlikely that the services installed in a building at the time of its construction will serve the users' needs for a long period of time. Indeed, it is probable that the services installation of a modern commercial building will be replaced several times during the functional life of the structure and fabric. For this reason it is common to come across the need to provide services installations on the basis of a retrofit approach. Here the services are installed after the building's initial construction and as such cannot readily be integrated with the fabric.

Principles of retrofitting

As with all operations involved in refurbishment work there is a scale of intervention, and this is also relevant to retrofitting of services. The scale is as follows:

Level 1: Leave everything as it is and do nothing except maintenance
Level 2: Add or remove items of equipment to improve performance of the system
Level 3: Refurbish the original systems and reuse them
Level 4: Replace certain items with similar and/or different ones
Level 5: Redesign and replace the whole system.

The choice of level to be adopted will depend on:

PART 3

- The condition of existing items and systems
- Whether the existing system conforms or is able to conform to current standards for services installations
- The performance requirements of the owner/occupier
- The ability to upgrade the system at a future date
- The size of the existing plant and how this encroaches on the usable floor area
- Whether the existing system complies with environmental policies.

There are many benefits of retrofitting both individual plant elements and whole systems. They include:

- Reduced energy costs
- Reduced maintenance requirements
- Greater reliability
- Improved internal environment
- Improved/modernised interior design.

The process of retrofitting

When designing a retrofit service installation, the first action that needs to be taken is an evaluation of the currently installed system. This will involve undertaking a survey of the systems and informing the client of the systems' condition. This needs to be undertaken before any design work starts and will enable recommendations for the new installation to be made. Original building drawings can be used to gather this information, but in many buildings that are to be refurbished, alterations may have been made during the life of the building and these may not have been documented. Therefore a desk study plus a visit to the building are recommended.

Once the survey has been undertaken, it is essential that the services engineer discusses the findings with the client. The focus of these discussions should be the scale of the proposed retrofit and the budget that the client has available, whether the building is to remain occupied during the works, and the client's performance requirements. From this information the services engineer will be able to propose a scheme which fits the client's requirements within an agreed budget.

If a complete retrofit is the preferred option, which is usually the case in buildings that are being extensively refurbished, access for the new system needs to be discussed with the architect. Floor to ceiling heights in existing buildings may restrict the possibility to house services under raised floors and above suspended ceilings, resulting in the need for different routes to be identified.

Services required in modern buildings

The main services required in modern commercial buildings are usually categorised as electrical or mechanical and may include those shown in Table 4.1.

There may be lifts installed that could be hydraulic or electrical, depending on the ability of the existing structure to accommodate supporting structures.

Table 4.1   The main services required in modern commercial buildings.

Electrical	Mechanical
Wiring for lighting and electrical sockets	Pipework for heating systems
Information technology cabling	Boiler installation
Smoke detectors and fire alarms	Ventilation and ductwork systems
Telephone cabling	Air-conditioning plant
	Plumbing installations (toilets, sinks etc.)
	Lift installations

Figure 4.56 shows a retrofitted air-conditioning system.

The requirements for building services have increased dramatically over the years, and it is a commonly held belief that the design of building services improves when the services engineer is consulted at an early stage in the initial design. This enables the building to be designed around the services, as opposed to the services installation being made to fit into an existing structure. The major difference between new build and refurbishment work is that the existing building will remain (either partially or wholly), and therefore the non-recommended approach to building services design will be the only option. Another problem that is specific to installing new building services into older buildings is that it may be difficult to comply with current legislation, especially Part L of the Building Regulations. The U values of the original structure may have to be improved, but also the performance of the heating system will have to be increased to counteract the poor thermal performance of the original structure.

Compliance with Part M of the Building Regulations, which deals with access for disabled people, also needs to be considered in the design, and this may mean that lifts need to be installed if it is not feasible to incorporate ramped access due

Figure 4.56
An air-conditioning system retrofitted to an existing building.

to the high gradient needed if a ramp is to be formed from the existing pavement to the building entrance.

The main techniques that can be used to provide lifts in existing buildings are:

- Stairway or stair carrier lifts, where guide rails are fixed to a stair wall and a seat or platform is used to lift people up the stairs. These are suitable for two-storey buildings only.
- Platform lifts, which comprise a freestanding self-contained lift unit, are easy to install and do not require a lift pit. Holes would be broken out through existing floors and the lift fixed into position.
- Through-floor lifts that require holes to be broken through the floors and a platform attached to two steel guides fixed to the existing walls. These are suitable for lifting one or two people or a person in a wheelchair. They may be manually or automatically controlled.

Reflective summary

- Modern buildings rely on sophisticated environmental control services and the ability to cope with IT infrastructure in order to be commercially competitive.
- It is unlikely that the services installed in a building at the time of its construction will serve the users' needs for a long period of time.
- It is probable that the services installation of a modern commercial building will be replaced several times during the functional life of the structure and fabric.
- It is common to come across the need to provide services installations on the basis of a retrofit approach.
- As with all operations involved in refurbishment work there is a scale of intervention, and this is also relevant to retrofitting of services.
- The level and amount of retrofitting will depend on a number of factors.
- When designing a retrofit service installation, the first action that needs to be taken is an evaluation of the currently installed system.

Review task

Consider the probable lifespans of the following building services installations and propose ways in which they may be upgraded during the life of the building:

- Passenger lifts
- Air-conditioning
- IT/data infrastructure.

4.7 | Remedying dampness

Introduction

■ After studying this section you should appreciate the broad principles involved in remedying dampness from a variety of sources.

■ You should be familiar with the main techniques adopted to treat rising damp, together with the various support measures that are necessary to ensure effective remediation.

■ You should also understand the methods of dealing with problems of penetrating moisture and condensation in buildings.

Overview

The mechanisms by which moisture enters a building have been examined in Section 3.8. This section attempts to provide an overview of the main options available to eradicate the damp problems. The technology associated with remedying dampness is simple and there are relatively few steps required to ensure an effective solution to most damp problems. The key to effective treatment is the accurate diagnosis of the cause of the dampness. In practice, misdiagnosis of damp problems is common and many remedial DPC installations have been undertaken when they have been unnecessary. Relatively few damp problems are the result of rising damp, and there are many reasons why electric moisture meters may give positive readings. The accurate interpretation of such readings is essential in drawing appropriate conclusions regarding an apparent damp problem. BRE and the manufacturers of moisture meters provide excellent guidance on their use and interpretation.

Remedial DPCs

The use of remedial DPCs to address the problems of rising damp is well established, and over the years has taken many forms. The purpose of a remedial DPC is to attempt to arrest the passage of moisture from the ground through the walls by inhibiting the natural process of capillary attraction. Although there have been many techniques attempted in the past, including atmospheric siphons and electro-osmotic systems, these have largely been replaced by two broad categories of DPC: physical and chemical. Since the use of electro-osmotic systems and atmospheric siphons has now been superseded by the more reliable and technically robust forms we shall not deal with them in detail here. However, they are worthy of consideration in overview since there are still examples of their use in existence. Thus the categories of remedial DPC can be summarised as follows:

- atmospheric siphons
- electro-osmotic systems
- physical DPCs
- chemical DPCs.

We shall now consider each of these in turn.

Atmospheric siphons

The principle behind the use of atmospheric siphons (or atmospheric tubes) is based on the increase of surface area of the wall suffering from dampness. The concept is quite simple. A series of porous ceramic tubes are inserted into the body of the affected wall. The hollow tubes provide an enlarged interface area between the damp wall and the surrounding air, thus promoting increased levels of evaporation from the surface. The intention is to remove excess moisture from the fabric of the wall by allowing improved evaporation. Tubes would be inserted into the wall at approximately 300 mm centres, each tube typically being 50 mm in diameter. The main problem with this approach is that the natural migration of soluble salts to the surface will block the pores of the tubes, and the level of evaporation of the moisture will be drastically reduced. Hence they have a limited effective lifespan.

Electro-osmotic systems

Although they were once used quite extensively, the slightly obscure systems electro-osmosis and atmospheric siphons have now been largely abandoned.

Electro-osmotic systems fall into two types: passive and active systems. They operate on the principle of creation of an electrical charge in the body of the wall that creates a repelling force to the charged ions of moisture that are attempting to rise through the wall. This effectively acts in the same way as a magnetic force, in which like charges repel and opposite charges attract. There is considerable scepticism about the effectiveness of these systems.

Physical DPCs

The remedial installation of a physical DPC is quite feasible provided that the wall construction is of an appropriate type. The use of this form of remedial DPC is recommended by BRE, as it represents a robust and reliable solution to the problem of rising damp. However, the installation must be carefully undertaken to ensure that a continuous barrier to moisture is created and that there is no adverse effect upon the structure of the wall. Physical DPCs are generally more expensive than chemical forms and can only be installed where an appropriate horizontal joint is present in the wall. A brick saw is used to open the joint in short lengths and a layer of impervious DPC material is carefully inserted, with adjacent sections overlapping to ensure that the barrier is continuous. As each short section is completed the joint is pointed to ensure that there is no risk of structural movement from closure of the joint. This is a laborious and painstaking process, and must be undertaken by skilled operatives.

PART 3

Figure 4.57
Remedial damp-proof
courses.

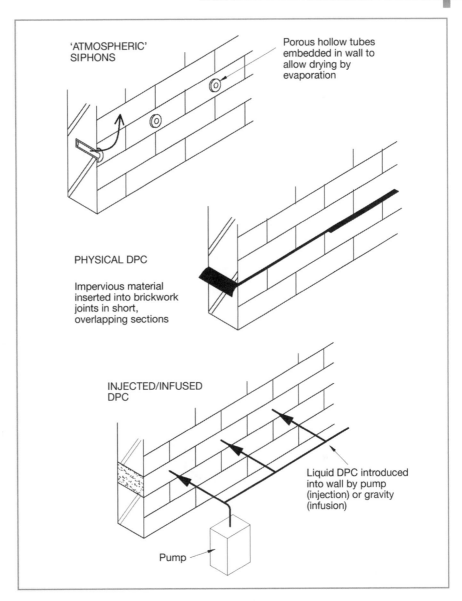

'ATMOSPHERIC'
SIPHONS

Porous hollow tubes
embedded in wall to
allow drying by
evaporation

PHYSICAL DPC

Impervious material
inserted into brickwork
joints in short,
overlapping sections

INJECTED/INFUSED
DPC

Liquid DPC introduced
into wall by pump
(injection) or gravity
(infusion)

Pump

Chemical DPCs

The vast majority of remedial DPC installations are based on the use of chemical DPCs. There are two broad categories of chemical DPC installation: infusion, which is based upon a chemical liquid being introduced into the wall under the action of gravity, and injection, which relies on the pressurised injection of the liquid using pumping equipment. Most systems work on the principle of either filling the pores within the wall's construction with a water-resistant material (pore fillers) or lining the pores with a non-wettable surface to reduce capillary attraction (pore liners).

The effectiveness of these systems relies in their effective penetration to the full depth of the wall and the provision of sufficient injection/infusion points to ensure complete coverage of the wall area.

Plaster renewal

The installation of a remedial DPC may not, in itself, cure the problems associated with historic occurrences of rising damp. The nature of the hygroscopic salts that are deposited as moisture rises through the walls is such that moisture problems will reoccur over time as the moisture from the air is absorbed by the salts. Hence the wall surface will still suffer from the presence of moisture and the visible symptoms of rising damp will persist. For this reason it is usual to remove the original plaster finish to a height of approximately 300 mm above the line of the rising damp in order to remove the hygroscopic salts. The wall should then be replastered to prevent any remaining salts in the body of the wall from migrating to the surface of the new plaster. For this reason, walls are often replastered at the lower levels using a base coat of sand and cement with added waterproofer, topped with a skim coat of plaster. Several proprietary plastering systems are available that fulfil the same function.

Figure 4.57 shows the different types of remedial damp-proof courses.

Reflective summary

- The technology associated with remedying dampness is simple and there are relatively few steps required to ensure an effective solution to most damp problems.
- The key to effective treatment is the accurate diagnosis of the cause of the dampness.
- In practice, misdiagnosis of damp problems Is common and many remedal DPC installations have been undertaken when they have been unnecessary.
- The use of remedial DPCs to address the problems of rising damp is well established, and over the years has taken many forms.

Review task

Outline the principles of the following remedial DPC systems, and suggest where their use may be most appropriate:
- atmospheric siphons
- electro-osmotic systems
- physical DPCs
- chemical DPCs.

4.8 | Repairs to masonry

Introduction

■ After studying this section you should have developed an understanding of how repairs to masonry can be achieved.
■ You should also be able to decide which system for rectification is most appropriate in given scenarios.

Overview

The principles associated with the failure of cavity wall ties have been dealt with in Section 3.4 and we shall not consider them again in detail here. The problem of corrosion of poorly galvanised or unprotected steel wall ties is widespread and it has been posited that the problem could eventually affect the majority of cavity-walled buildings constructed prior to the 1980s. In significantly older properties the problems are likely to have been recognised and treated already and the symptoms of such failure have already been noted. For many newer properties the process of 'quiet failure' may still be ongoing and, as yet, unrecognised. Hence the extent of cavity wall tie renewal that is still required in the UK is significant.

Remedial cavity wall ties

Replacement of damaged or defective ties will ordinarily be with non-ferrous materials such as stainless steel or copper. In practice the use of stainless steel is almost ubiquitous, although the use of plastic forms of remedial tie is on the increase. Two basic principles are adopted for fixing the remedial ties (Figure 4.58) into the affected wall:

■ Expanding bolts
■ Resin or grout fixing

Figure 4.58
Remedial expanding bolt and resin fixed ties (© Ancon Building Products).

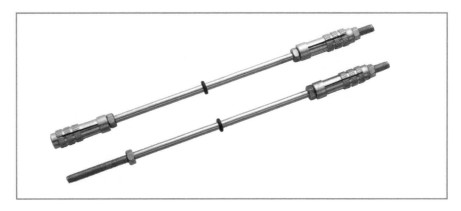

PART 3

In both instances the remedial tie is inserted into the body of the wall from the outside. Holes will be drilled through the body of the bricks on the outer leaf and into the brick or block inner leaf to an appropriate depth. The remedial tie is then installed and either tightened to expand the external sleeve of an expanding bolt type or grouted into position with an epoxy resin to secure both the inner and outer ends of the tie. In all cases it is recommended that the existing corroded tie is removed from the outer leaf in order to reduce the potential for continued corrosion, lamination and deterioration of the outer leaf by cracking.

From an insertion point of view it should also be remembered that the replacement tie is inserted into the body of the brick/block rather than into the mortar joint, where the original tie would have been bedded. However, the spacing of the new ties will mimic the recommended spacing of new cavity wall ties as far as possible.

Another key point is that when selecting a replacement tie system the different ways in which the ties can be secured in the body of the wall may be more or less appropriate in given situations. Expanding ends to the ties are quite acceptable where the wall material has high strength, such as dense brick, but where less strength is available, as in older bricks or lightweight blockwork for example, a resin-bonded end to the tie is far more typical. Such ties will normally be anchored in epoxy resin.

A range of BRE Digests are prime references for identification and treatment, and the appropriate publications are Digests 329 and 401.

In conditions of low exposure the use of an injected cavity wall foam insulant to effectively bond the inner and outer leaves together has been used with some success. This is far less common than the other more traditional methods and must be installed by an approved, specialist contractor.

Dealing with cracking in walls

It is important to recognise that cracking in masonry is generally a symptom of some other defect arising from problems above or below ground. The main problems that might give rise to the appearance of cracks have been considered in Chapter 3, and many of the remedial measures are self-explanatory. For example, the obvious treatment for a failed beam or lintel may be to simply replace the failed item with a suitable new unit. Other aspects, such as treatment of settlement, have been considered specifically in Section 4.3, dealing with underpinning. However, the treatment of cracking should also be considered as a separate issue, since the presence of the crack itself may lead to other consequential issues such as damp penetration and further deterioration of the building fabric. In general, cracks that are below 5 mm in thickness are considered 'safe'. Those above this limit will normally require treatment to the affected area.

The level of intervention that will be required depends upon the cause of the cracking and its severity. Treatments can be considered within a range of broad categories.

Non-structural cracking, such as thermal movement cracking, may be dealt with by simply repointing the affected area. This generally relies on the use of a relatively 'soft' mortar to allow slight movement to take place without cracking.

Structural cracking, such as settlement cracking or cracking arising from failure of lintels, overloaded wall sections etc., requires more significant treatment. It should be noted that these treatments will be used in conjunction with the necessary repairs to arrest the structural movement (i.e. underpinning, replacement of beams/lintels), and will not be used in isolation. Treatments for structural cracking may include:

■ Resin grout injection. Here, an epoxy resin is injected into the crack under pressure to achieve a physical bond between the sides of the crack.
■ Removal and rebuilding of affected wall sections.
■ Stitching of the crack using steel straps or rods.

Deformation in walls

The extent to which walls will deform is closely linked to the degree to which they are provided with appropriate lateral restraint.

In buildings of considerable age it is not uncommon for external and internal walls to suffer deformation due to oblique or lateral loads and the absence of sufficient lateral restraint within the structure. This may also be a symptom of overloaded wall sections that have started to fail through buckling. In order to arrest further deformation it is common to adopt the use of steel restraint ties that secure the external walls to internal floors and walls in order to achieve a degree of lateral restraint. In some instances these restraint fixings will pass through the entire building, effectively securing opposite external elevations to each other.

Reflective summary

■ The extent of cavity wall tie renewal that is still required in the UK is significant.
■ Replacement of damaged or defective ties will ordinarily be with non-ferrous materials such as stainless steel or copper.
■ In practice, the use of stainless steel is almost ubiquitous, although the use of plastic forms of remedial tie is on the increase.
■ A key point when selecting a replacement tie system is that the different ways in which the ties can be secured in the body of the wall may be more or less appropriate in given situations.
■ It is important to recognise that cracking in masonry is generally a symptom of some other defect arising from problems above or below ground.
■ Non-structural cracking can be rectified by simply repointing the affected area.
■ Structural cracking requires more significant treatment.
■ In buildings of considerable age it is not uncommon for external and internal walls to suffer deformation due to oblique or lateral loads and the absence of sufficient lateral restraint within the structure.

Review task

Describe the two basic principles that can be adopted for fixing remedial wall ties into an affected wall.

PART 3

4.9 | Treatment of timber defects

Introduction

- After studying this section you should have developed an understanding of how defects in timber can be treated effectively, and how prevention of a reoccurrence of the problem can be incorporated into the affected element of the building.
- You should also be aware of how to find more detailed information on the treatments for timber defects.

Overview

In the context of timber defects we are generally concerned with softwoods, which are less durable than hardwoods.

The key element in the durability of softwood is the degree to which it is able to cope with moisture. BS 1186 refers to timber moisture content and advises on desirable levels of moisture. Typically, a softwood in use internally, as skirting board for example, would have around 8–10 per cent moisture content *in situ*. However, timber is *hygroscopic* and it will absorb moisture from the surrounding air. This is a particular concern when dealing with fungal attack in timber.

Fungal rot development

When treating dry rot it is essential that support measures are put into place to ensure that the environment that resulted in the initial outbreak is controlled.

The key factor in the development of fungal rot is the moisture level within the timber. A consistent level of moisture of at least 20 per cent is required to encourage the development of dry rot, and typically over 22 per cent for wet rot.

The main information source for remedial procedures and identification are the BRE Digests 299 for Dry Rot and 345 for Wet Rot. These divide remedial treatment into *primary* and *secondary* measures.

Remedial action for dry rot

There have been recent developments in the treatment of dry rot in buildings, with some specialist firms now advising that the traditional, destructive remedies for dry rot should be replaced with alternatives. The alteration of environmental conditions around the outbreak, reducing moisture levels, increasing temperature and providing good levels of ventilation, has produced good results. However, this form of treatment, although advocated by notable specialist firms, is hindered by professional indemnity restrictions. Hence more traditional methods will be most common in all but buildings of particular architectural interest for some time to come.

Traditional treatment will typically consist of the following:

- Eliminate moisture sources
- Attempt to dry the affected areas rapidly
- Remove rotted wood to within 450 mm of visible outbreak
- Irrigate surrounding brickwork etc. with fungicide
- Adopt support measures, such as increased ventilation levels.

Remedial action for wet rots

Wet rot is far more easily treated than dry rot, but again it is important to recognise the need to locate and eliminate the source of moisture that caused the problem in the first place. There are several forms of brown rot within this category, some of which may be confused with dry rot. The expensive measures which are adopted for the eradication of dry rot need not be employed for these forms, which are less extensive in terms of the damage they cause. Common forms found in buildings are: cellar fungus (*Corniophora puteana*), *Poria* species, white rots and other lesser known varieties. Similar measures are required for the eradication of these forms; hence they will be considered together.

The nature of wet rot attack is far less devastating than that of dry rot. However, treatment is still necessary in most cases. The treatment of all wet rots is essentially similar and is summarised as follows:

- Eliminate the source of moisture
- Promote rapid drying of the affected areas
- Remove the rotted wood and replace with preservative-treated wood
- Adopt support measures.

Insect infestation

BRE Digests 307 and 327 supply some useful information. One of the aspects of recognition and treatment of insect infestation is the Categorisation class in Digest 307. This gives three classes of insect: Category A, which requires insecticidal treatment; Category B, which requires treatment only to remove the conditions causing wood rot; and Category C, which requires no treatment at all.

The fact that the insect in grub form (the longest time span in its life cycle) does most of the damage will sometimes influence the method of treatment suggested.

Remedial action for insect attack

It is assumed that remedial action is only to be considered for insects falling into Category A, since those falling into Category B will be treated for their fungal attack rather than the infestation of insects. It is important to correctly identify the insect type to enable it to be correctly categorised. All insects falling within Category A will, however, be treated in essentially the same manner. The procedures for remedial treatment may typically include the following stages:

PART 3

- Identify the insect
- If within Category A, determine whether the infestation is active or dormant
- If the damage is linked to fungal attack, treat the fungal attack as previously described
- Assess the significance of damage (20 holes per 100 mm run = severe attack)
- Apply suitable remedial treatment.

Reflective summary

- In the context of timber defects we are generally concerned with softwoods, which are less durable than hardwoods.
- The main information sources for remedial procedures and identification are BRE Digests 299 for dry rot and 345 for wet rot.
- There have been recent developments in the treatment of dry rot in buildings, with some specialist firms now advising that the traditional, destructive remedies for dry rot be replaced with alternatives.
- However, more traditional methods will be most common in all but buildings of particular architectural interest for some time to come.
- Wet rot is far more easily treated than dry rot, but again it is important to recognise the need to locate and eliminate the source of moisture that caused the problem in the first place.
- One of the aspects of recognition and treatment of insect infestation is the Categorisation class in BRE Digest 307.

Review task

Refer to the relevant guidance mentioned above and produce a detailed plan of how to treat dry rot.

Management of maintenance and refurbishment

The management of refurbishment work

After studying this chapter you should be able to:

Explain the issues related to the design of refurbishment work

Demonstrate an understanding of how the design of new build works differs from that of refurbishment

Identify the need for client input into the design process for refurbishment work

Outline issues related to the design and management of refurbishment projects if buildings are to remain occupied

Detail what is required in the appraisal of a building where refurbishment is proposed

Identify the different procurement methods that can be used in refurbishment projects and recommend suitable approaches for different types of contract

List and discuss the approaches to the management of refurbishment contracts, which if employed correctly should facilitate the efficient management of the works

This chapter contains the following sections:

5.1   Management of design
5.2   Procurement and management of construction

- CIOB Construction Paper No. 66 1996: Characteristics and difficulties associated with refurbishment
- CIRIA Report 113: A guide to the management of building refurbishment
- RICS Guidance Note on Planned Preventive Maintenance
- RICS Guidance Note on Maintenance Policy and Planning
- Standard Maintenance Descriptions (RICS Research Paper)

5.1 | Management of design

Introduction

- After studying this section you should have developed an understanding of the issues that relate to the design of refurbishment schemes and of the differences from the design of new building work.
- You should also be able to explain why some clients will choose to keep a building occupied during construction work, and the problems associated with this decision.
- You should be able to discuss the issues that the client needs to address when commissioning refurbishment work, and factors that are specific to refurbishment work from a design perspective.
- In addition, you should have developed a very real understanding of the need for a detailed appraisal of any building that is going to be refurbished, even if there is a high cost associated with this activity.

Overview

Refurbishment design as opposed to new build design poses a number of specific problems. In new build design it is common practice to start with a blank sheet of paper and build up the design, but with refurbishment design it may not be feasible to adopt this approach, and may be easier to look at specific areas within the building on an individual basis and then link these areas to produce the overall design. In the case of refurbishment there is already an existing asset, which has a value, and any design for new work needs to ensure that the existing asset value increases or at least remains static.

An additional problem that can influence the design of refurbishment work is that the building may need to remain occupied while work is ongoing. This may be because the earning capacity of a company needs to be maintained during the works and there is no other suitable accommodation, or because the building needs to be maintained for public use (e.g. a railway station).

Client responsibilities

It is important that clients brief designers at length before any design work is undertaken. One of the major problems in refurbishment is that clients do not make it absolutely clear what it is they require from the proposed refurbishment scheme. They do not identify the required performance level of the intended building, and give limited information regarding space requirements. This then leads to a design which dictates cost and construction work that when complete will leave a disillusioned client. These early discussions are especially important if the building is to remain occupied during the works, because occupation during the works may lead to a limitation in design, and clients need to be fully aware of this. Ideally the client should appoint an experienced representative who is knowledgeable about the client's working practices and requirements. This representative should take responsibility for the management of the design process. This is especially important from a health and safety perspective, as the client has responsibilities for health and safety under the Construction (Design and Management) Regulations. An experienced representative should also make the client aware of the longer lead-in time required for completed designs for refurbishment work, and give advice as to which contract would be the best option for the proposed project. A contract that allows for an overlap between design and construction would be the best option in some cases, although this will lead to uncertainty of final cost to the client. The client also needs to be made aware of the fact that a major feature of refurbishment contracts is the discovery of 'unknowns' throughout the duration of the contract. These may slow down work significantly and incur additional costs. The client needs to be strongly advised to set monies aside to allow for these contingencies. The nature of these unknowns and the extent of the problems that they can cause will be different for every building, but they include:

The design options for layouts in buildings to be refurbished are likely to be more restricted if the building is to remain occupied.

- The discovery of dangerous materials that were used in the initial construction of the building (e.g. asbestos and lead). These will require removal by specialists.
- The discovery of dry rot, wet rot and woodworm in timber.
- The discovery of damage to main structural elements such as foundations. If this occurs underpinning may be necessary.
- Services that were believed to be obsolete but which are still live.

The client may, however, qualify for funding from grants, which tend to be available for refurbishment work but less so for new build work. The availability of these grants is often linked to the design of the scheme, and higher levels of funding could be obtained via grants if the proposed building will achieve a greater perceived status than the original building.

Finally, it could be beneficial for the client to undertake a risk assessment based on a number of proposals for schemes of work to the existing building. This should assess the risks from cost and constructability perspectives, and clients may wish to choose a scheme that may not achieve the exact building performance levels required, but in which there will be less risk to themselves from making this choice.

Design issues

Before any initial design is produced, a detailed appraisal of the building is required. However if a client is trying to keep costs to a minimum it may be decided to procure a less detailed appraisal. The aim of a detailed appraisal of an existing building is to try to minimise the number of unknowns that may be uncovered during building work. Therefore a higher level of investigation work at the appraisal stage may prove more expensive initially but be very effective in keeping costs to the original budget in the long term.

Part of the appraisal process may be to look at the original building drawings if they are available, but this is no substitute for on-site investigations. For example, the original drawings may be available, but extensive works might have been carried out over the life of the building which have not been recorded. The drawings may bear little or no resemblance to what is actually in the building. Another important part of a detailed appraisal is to test existing services to see whether they are still live and can be used as part of the new scheme, although this is unlikely due to rapid developments in building services and increasing requirements from clients regarding services performance. Before any design work is undertaken, an appraisal of the existing building is of paramount importance, but it is also essential that buildings that surround the proposed site have some appraisal, as the state of these buildings may affect the work that can be done.

Information required before an appraisal is carried out

- Before the commencement of an appraisal, the appraisor should be very aware of the intention and expectations of the client.
- The level and standard of the scheme should be agreed by the client and designer.
- The relationship and balance between the final building value and required building performance need to be established.
- The building owner may wish to restore the building to its original state due to damage being caused accidentally.
- However, in most cases the building owner may want to include improvements or modifications.
- The extent to which demolition is possible needs to be established. This may be dictated by adjacent properties or by planning restrictions.
- An initial idea of the proposed budget for the refurbishment is essential, as this will obviously affect the extent of a scheme.

If all of these have been addressed, this should ensure that all risks have been identified, be they health and safety risks, financial risks or planning risks.

The design team and design process

Demolition is required to some extent on most refurbishment contracts. It is identified by the HSE as one of the most dangerous activities on-site.

It is a commonly held belief that if all the parties involved in the design of buildings worked together at an earlier stage, then the completed design would be of much better quality. This is even more the case for refurbishment work, and early collaboration of the design team is essential. Refurbishment work is inherently more dangerous than new build work due to the requirement for dangerous activities such as demolition, and the designers have a responsibility for health and safety under the Construction (Design and Management) Regulations. Therefore, for that reason alone it would be good practice for them to discuss potential high-risk activities and try to reduce the risks through a well-considered design scheme.

One of the main keys to the design of a successful refurbishment scheme is to build-in flexibility. This flexibility should allow for solutions to problems that occur due to unknowns to be developed quickly and easily with little or no cost implication. The proposed design also needs to focus on the detection of what is present in the existing building, and making best use of these features.

One of the major problems in the design of refurbishment schemes is compliance with current Building Regulations. For example:

- Buildings need to comply with Part M and allow for disabled access. This may be very difficult to achieve if access to the existing building is via a number of steep sets of steps. Will it be possible to form a ramp of a suitable gradient into the building? If not, will existing floor levels need changing?
- The changes to Part L of the Regulations require rigorous testing of air leakage from the completed refurbishment scheme. How can it be guaranteed that these levels can be achieved?

Other issues that need to be carefully evaluated are:

- The interaction between the existing building and proposed new work. Differential settlement can be a major problem in refurbishment work and any design needs to take account of this.
- The 'squareness' of the existing building. Most buildings today are designed as square schemes, where most elements of a building are designed to be at 90 degrees to other elements either horizontally or vertically. However, in most older buildings this is not the case, and rooms can be severely out of square and walls significantly out of plumb. Design solutions need to be developed to eradicate or make best use of these problems.
- Most older buildings will have been designed using imperial measurements, and this may mean that modern standardised components will not 'fit' into older building fabrics. The designer must decide whether or not the use of one-off building components is feasible from a cost perspective, or whether it will be better value to demolish a particular area of the building and build new so that off-the-shelf components can be used.

Finally, the proposed scheme may require the renovation of existing features. The designers may need to consult old building construction textbooks in order to determine how the renovations can be achieved, but even if this is done there may be a problem finding people who possess the skills required in order to undertake the work. If this is the case, a design solution needs to be developed before any work is specified.

PART 4

Review task

Produce a checklist that could be utilised by a designer for the appraisal of an existing building.

5.2 Procurement and management of construction

Introduction

- After studying this section you should have developed an awareness of the different forms of procurement that are possible for refurbishment work and the advantages and disadvantages of these methods to both the client and the contractor.
- You should also be able to discuss the issues that need to be addressed by the site management team in order to facilitate the efficient management of refurbishment contracts during the construction phase.

Overview

The forms of contract used for refurbishment contracts are basically the same as for new build work. However, some procurement techniques may prove more beneficial for the client, the contractor or both. In any contractual arrangement it

Figure 5.1
Primary and secondary objectives for construction work.

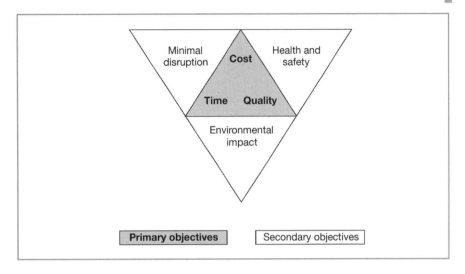

is better to aim for win–win situations, and in order to achieve this the procurement option chosen should be carefully considered.

The basic control mechanisms for cost, duration and quality of the works that are used on new build contracts should also be used for refurbishment contracts, but they may take slightly different forms. Figure 5.1 illustrates how the primary factors of time, cost and quality can be expanded to include secondary factors of health and safety, environmental issues and disruption factors, which may be more difficult to control on refurbishment contracts than on new build sites. Indeed, the secondary factors can become more important than the primary factors when trying to control refurbishment work. This is even more likely if the building is to remain occupied throughout the building work.

Disruption factors include noise, dust, fumes and dirt produced on construction sites.

Procurement options

There are many possible options:

- *Selective or open competitive tendering*
  Although this form of procurement is commonly used, it can create problems for the client. Tender prices and contract durations are submitted by contractors based on drawings and bills of quantities. This works reasonably well for new build work, as it is easy to measure the quantities of materials required and to determine labour level requirements, as all the information is given in this initial documentation. The discovery of unknowns is rare and usually restricted to groundwork elements of the contract. However, in refurbishment work the number of potential unknowns that will be discovered during the works could be huge, and these will not have been priced by contractors at the initial tender stage. These extras will have to be dealt with as variations to the contract and as there will generally be no bill quoted activity to compare the works against, they may have to be undertaken as daywork

items that can be costly to the client. From a contractor's perspective, however, these variations due to extras can increase profit if dealt with correctly.

- *Two-stage tendering*
  This method is a variation of the above, but is better suited to refurbishment contracts. Initially, competitive bids are invited from contractors for works that can be easily detailed and specified. Once the contract has been awarded, the cost of additional works required will be negotiated using bill rates that have been determined by predicting possible work requirements. The advantage of this method is that there is a competitive element followed by collaboration in order to reduce the cost of extras.

- *Drawings and specification*
  Because of the difficulty in producing a bill of quantities for refurbishment work, the use of drawings supplemented by a technical specification is often used as the basis for pricing by contractors. The main problem with this method is that there is a significant risk to the contractor. Building work requirements may arise that have not been specified on the drawings, and these can incur additional costs that may have to be borne by the contractor. The major risk to the client in this situation is that the contractor may not be able to complete the works due to the lack of funding, and may pull off the site. A different contractor may then need to be employed to complete the works, and they will demand a significantly higher level of funding to do this. Additionally, the use of drawings and specification may lead to a reduction in competition, because each contractor could interpret the specification differently. The lowest price contractor may have misinterpreted what the client's requirements are, and the client may be left with a building that does not meet its performance requirements.

- *Negotiation*
  Negotiation is when the cost of all works are negotiated by the client and contractor as and when required. This means that contractors should always be satisfied with how much they are getting paid and should lead to improved quality of the works. However, the problem for the client is that there is no competition, and the cost of the works will tend to be significantly higher than in competitive tendering. Negotiation tends to be the preferred method when construction work is phased, i.e. the contractor successfully completes the works on Phase 1 to the standard required by the client, and then negotiates costs on additional phases based on bill rates for Phase 1.

- *Partnering*
  This is similar to negotiation, where a contractor is awarded a number of similar contracts based on a competitive tender price for an initial contract. The principle of partnering is that a win–win situation should occur: the client gets a building of a suitable quality at an acceptable price, and the contractor makes a

reasonable profit with reduced risk. However, since all refurbishment contracts are very different, it can be difficult to state categorically that a contractor will perform to the same level on each contract.

- *Cost reimbursement and fee*
  In this method, the contractor is paid for the works undertaken and is also paid a fee for managing the work. This is excellent for the contractor, but it means that they do not need to be careful and procure the lowest costs possible for the works. There is therefore a significant financial risk to the client. The benefit of using this method for the client is that there is a reduced lead-in time before work can commence. A detailed appraisal may not be required, and detailed drawings or a bill of quantities do not need to be produced. Work can start immediately, and the design will be ongoing as work proceeds, which allows for the additional benefit of design flexibility, where the most can be made of existing features that may not have been apparent before work commenced.

- *Design and build*
  In design and build schemes, the contractor is responsible for the design and construction of a building. The client needs to brief the contractor in detail about its requirements, and may fund the cost of an appraisal by the contractor. The contractor can then undertake the investigations that it deems to be necessary in order for its design to develop. On paper, this is the best approach, as the client will be able to compare a number of schemes and evaluate their preferred option based on the design, the cost and the contract duration. Any additional costs will generally be borne by the contractor. However, the disadvantage is that if there are a large number of unknowns that the contractor did not expect to find, the contractor may reduce the specification of the works to try to reduce overall costs. This may lead to a reduced quality building being constructed that does not meet the client's performance requirements.

Although, on paper, design and build may seem to be the best procurement option for clients, it is not commonly used in refurbishment contracts.

To summarise, there is no procurement method that does not have any risks for the client, the contractor or both. The decision to use a particular method will depend on:

- Client preference
- Client advisor preference
- Size of contract
- Type of refurbishment contract (minor, major, complete etc.)
- Quality of completed building work required by the client
- Client budget flexibility
- How quickly the client wants the completed building
- Whether the building to be refurbished is to remain occupied during the works
- A prediction of the potential number of unknowns.

PART 4

Management of construction

As has previously been stated, the control mechanisms that contractors use for the management of refurbishment contracts are the same as for new build work. However, there are subtle differences and some aspects will take on greater relevance and importance than others. The major difference between refurbishment and new build from a construction perspective is that, in new build work, methods of construction are reasonably standard across the industry. For example, if a new multi-storey steel frame is to be constructed, the sequence of construction will be as follows:

- Excavate foundations
- Fix holding-down bolts
- Concrete foundations
- Connect ground floor columns to the holding-down bolts
- Erect the steel frame.

There is no variation on this method, and every contractor would use the same method for the same frame.

However, in refurbishment, because the works can be very fragmented, different contractors could undertake the works in different sequences to produce the same finished product. The success of a refurbishment contract is very much linked to the ability, knowledge and conscientiousness of the site team.

An initial requirement of the site team is to establish relationships with the client. This is especially important if the building is to remain occupied. The client needs to be very aware of the planned progress of the work and the implications for the occupiers of the building. Staff may need to move offices on a number of occasions to help facilitate the works and they need to know this well in advance. There may be periods during the works when noise restrictions will apply and the contractor needs to know these periods well in advance to reduce delays in progress. A further factor that needs to be communicated to the client is that of progress. In new build schemes, rapid progress is seen even on a weekly basis and this reassures clients that their money is being put to good use. Progress of refurbishment contracts can be difficult for a client to detect, as there could be a large amount of enabling work required. The client may become concerned that refurbishment was not the correct route to take and will need reassurance from the contractor.

Schedules of conditions

It is essential that schedules of conditions include photographs.

From a contractor's perspective, it is very important that a schedule of conditions is produced before any work commences. The schedule should detail the condition of the building, and should include date-coded photographs of any elements of the building that are planned to remain in the proposed scheme. The purpose of this is to prevent a client from accusing the contractor of damaging existing features if this is not the case. Quite often, elements of the building may appear to be in reasonable condition when the building is in a state of disrepair, but when new work is completed around these elements the extent of the original damage becomes apparent. The client could quite easily be mistaken in accusing the contractor of damage, but the only way to prove this is to refer back to the schedule of conditions. Although there is a cost attached to the preparation of this schedule, in the long term it can prove extremely beneficial to the contractor. After the schedule has been produced, it needs to be signed by the client and/or designer as being a true record of the state of the original building.

Choice of site personnel

As has already been stated, the site personnel on refurbishment contracts need to have highly developed communication skills. There is a great deal more interaction with the client and designers in refurbishment work as opposed to new build work. On new build contracts it is highly likely that the designers will only visit the site once or twice a month, and the client less so. Because of the nature of refurbishment work and the need for rapid decisions due to the detection of unknowns, the designers will often visit a lot more frequently. Clients may choose not to visit any more frequently if the building is unoccupied during the works, but if the building is occupied, in effect they are on-site for the whole duration of the contract.

Site paperwork and quality systems

Variations and extras are highly likely on refurbishment contracts, and as such the contractor's quality systems need to be highly developed in order to secure funding for all additional works. If new details are provided by the designers on-site, then their receipt needs to be logged immediately. Drawings tend to be revised more regularly on refurbishment schemes and all parties affected by changes to drawings need to be informed to avoid building from superseded drawings. The quality system needs to allow for the rapid decision-making process that occurs on refurbishment contracts, and site paperwork needs to be carefully filed and stored so that it can be easily found if and when required.

Management of disruption

During refurbishment work, the main factors that can cause disruption are:

- Noise
- Dust
- Fumes
- Dirt.

PART 4

Although these factors can be present during new build work, it is more likely that they will be encountered during refurbishment, and at greater levels. This is due to the fact that most refurbishment contracts will include demolition, which causes a great deal of noise and dust. These problems will need greater control if the building is to remain occupied during the works, but even if the building is unoccupied, neighbours need to be considered.

Noise can be controlled by using barriers and by restricting noisy work to hours where occupants will not be affected. Dust can be controlled using water spraying, and fumes can be reduced by providing adequate ventilation and using machines that do not emit fumes. The amount of dirt produced on the site can be controlled by cleaning pedestrian areas as thoroughly and regularly as necessary.

However, the need for providing this extra protection from disruption will lead to additional costs.

Programming and interface management

Programming of refurbishment work can be much more complex than programming new build work. In new build contracts the activities required tend to be fewer and take longer to undertake, which enables the production of a programme which has limited activities and can be presented in a simple manner. However, in refurbishment the work tends to be very fragmented, with a lot of small jobs required in order for more major works to be undertaken. For example, a new steel column may be required to give additional structural stability. This will require excavation of a small pit for a foundation, fixing one set of holding-down bolts, concreting-in the one set of bolts, and fixing the column top and bottom. Each of these activities may only take an hour or so, and there could be a number of similar instances. If all these individual activities were to be put on a programme then the detail would be immense and make the programme over-complicated. The problem with overcomplicated programmes is that there is a tendency for site staff to ignore them!

Reliance on the site team to 'remember' to do all these activities could then become the norm, and there is no real mechanism for control of the construction process.

Another major potential problem on refurbishment projects is that of interface management. Although this can be a problem on new build sites, most tradesmen will have an understanding of how their area of work interfaces with other areas of work. For example, bricklayers will understand that in order for them to build a wall, they need a foundation to build on, and they will know that first fixing occurs when they have built the wall. However, in refurbishment work this can be far from clear, and it is the responsibility of the site team to communicate this information very clearly to their own labour or subcontractors. Some of these potential problems can be avoided if work packages are carefully compiled. Rather than just splitting up a bill of quantities into what would be traditional subcontracts, it may be better for the contractor to directly employ labour that has the skills to undertake small areas of specialist work in order to avoid interface prob-

> Work packages are sections of work undertaken by different subcontractors.

Figure 5.2
$1/4 : 1/3$ rule for different building scenarios.

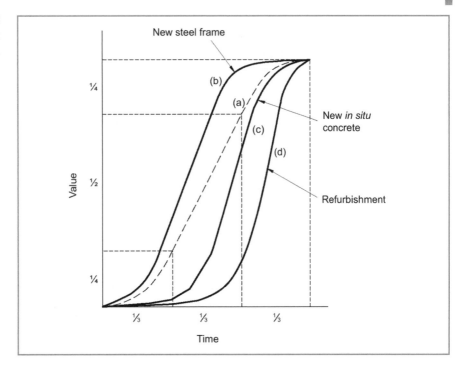

lems. Given the previous example, it would not be necessary to employ a steel-work subcontractor to install one column, provided the contractor's labour has the ability to undertake this work.

Control of cash flow

The cash flow requirements for construction projects can be predicted by the production of a cash flow diagram based on the programme. Given the problems detailed previously, this can be difficult to achieve with any great level of accuracy. Some contractors predict cash flow based on the rule, where they assume that a quarter of the contract value will be expended in the first third of the contract duration, half will be spent in the middle third of the duration and the final quarter in the final third of the contract. This model works reasonably well for new build contracts, but for refurbishment work it could be totally unsuitable. Figure 5.2 shows graphically the model rule. It also shows how this could be inaccurate when comparing the construction of a new *in situ* concrete building and a new steel frame building. In a steel frame building, because the steel is constructed quickly at the beginning of the contract, a lot of money will be spent early on in the works, but in the case of *in situ* concrete the construction of the frame will proceed more slowly, and therefore the cash flow will be slower even if the overall cost is the same for both. To counter this problem, contractors may develop a series of model S curves, as shown in Figure 5.2, built up from previous experience for different contract types. However, in refurbishment work there is no model to compare against, as every contract will be very different. The S curve as shown illustrates that on refurbishment contracts the cash flow initially will be very slow, as low cost activities may need to be carried out in order to enable more substantial activities to be undertaken.

Access, storage and accommodation

Access, storage and accommodation are easy to plan on open sites where there is plenty of space around the building. This is the same for refurbishment and new build contracts. However, large numbers of refurbishment contracts are in town and city centres, where there is very limited (if any) space around the site. If this is the case then these issues need to be very seriously considered at pre-tender and pre-contract stage, and will be more problematical if the building is to remain occupied.

■ *Access*: access for materials deliveries may be very limited, and all materials may need to be physically carried into the site by labourers. The use of cranes and hoists may be restricted due to the nature of the works. This is very labour-intensive and the amount of labour allocated to the site will be dictated by this. If materials are required that are too heavy to manhandle, but installation of a permanent crane is not feasible, then mobile cranes may be used if and when required. If mobile cranes are used, closure of roads may be necessary and this will only be allowed if the contractor has applied to the Local Authority for a road closure. It can take a significant length of time to be given permission for these, and therefore if key dates are missed it could severely delay the contract if the materials cannot be delivered. In occupied buildings, routes have to be clearly defined for access into the building, as it is uncommon for building workers and building occupiers to use the same entrances.

The term *double handling* can mean moving twice, but may mean moving again and again and again!

■ *Storage*: on most refurbishment contracts there is little space for storage of materials. This will result in the need for smaller and more regular materials deliveries, which will increase the unit cost of the materials. If this is not planned correctly, it will result in a large amount of double handling of materials, which will increase the cost of the contract.

■ *Accommodation*: site staff and operatives require accommodation throughout the duration of any contract. When undertaking new build work in open areas, most contractors will use portable buildings for this. However, this may not be possible on confined refurbishment contracts. Alternatives for providing accommodation include:
  – Stacking of portable buildings on a scaffold erected up the side of the building. There is a fire risk in this situation: if a fire develops in a cabin at the bottom of the stack, this will quickly spread to all the other cabins.
  – Using an area of the building as accommodation. This is entirely feasible, but may mean that frequent movements of accommodation occur. Offices may be situated in an area where no work is being undertaken, and then move to a different area when work is required in the previous area. This may need

to occur on numerous occasions, and there is a cost implication. This may not be feasible if the building is to remain occupied, due to lack of space.

– Renting offices away from the site itself. If neither of the above is feasible, then the contractor may need to rent office and accommodation space in other buildings. This will be satisfactory if accommodation can be found near to the site so that walking distances are kept to a minimum. However, it can be very expensive, depending on the rental costs in the area, and usually some refurbishment work of the offices that have been rented will be required at the end of the contract.

All of these options have a cost implication, and this needs to be considered by the contractor early on in the planning process.

Health and safety

Health and safety are major issues on every building contract, and Health and Safety Plans and Method Statements need to be produced to ensure that works are carried out safely. However, on refurbishment contracts these Method Statements may be more difficult to produce due to the number of unknowns that the contractor could find. For example, it is extremely likely that asbestos will be discovered during refurbishment work. If this is the case then a specialist subcontractor needs to be called in immediately to remove it. Demolition is classed by the Health and Safety at Work Act as one of the most dangerous activities to undertake, and on virtually every refurbishment contract there will be demolition to a certain extent. This work needs to be very well planned before it is undertaken. Even though many of these issues may not be predictable on refurbishment contracts, a process for dealing with them will be established by the contractor.

Fire is another potential problem, and again even more so when the building remains occupied. Very serious precautions need to be employed to prevent any spread of fire in the existing building, because the planned fire safety systems for the scheme will not be in place during the works.

Workforce

The managerial and communication skills required of refurbishment construction managers have been discussed previously. A high level of technical knowledge is also essential in both operative and managerial staff. The main difference between new build and refurbishment contracts is that, on new build sites, gang sizes may be reasonably large and therefore a subcontractor can justify the employment of a full-time supervisor on the site. On refurbishment contracts, gang sizes may be considerably smaller and the subcontractor may deem it not cost-effective to supply supervision. If this is the case, the main contractor's supervisory staff will be expected to manage subcontract labour, and this can be very difficult.

Setting out

Setting out of new build work is a relatively easy task. New build designs are based on square grids, and work basically starts from a 'blank canvas'. It is easy to establish datums and horizontal setting out lines because sights are not impeded by parts of the existing building. Profile boards can be utilised to enable the establishment of both line and level.

In refurbishment work, establishing datums and gridlines can be very difficult because it will be difficult to sight through a gridline as it is likely to be obscured by part of the building. It may be that the only way to establish gridlines is by using plumblines. Holes would need to be broken through each floor at a number of points, plumblines dropped through and gridlines marked on the existing floor after taking measurements from the plumbline. An additional problem with existing buildings is that they may not be square. The new scheme will almost certainly be designed using gridlines that are at 90 degrees to each other, and the designer will need to find a solution to this problem.

Datums will need to be established on every existing floor level, and the only way to achieve this will be to undertake flying levels up and down existing staircases, which can be difficult.

New services installations

Integration of services can be problematical on most construction sites, and the contractor needs to ensure that somebody is taking the lead role in facilitating this. The problem could be compounded further in refurbishment contracts because of:

- Limited existing floor-to-ceiling heights, reducing the space available for services.
- Possible connection of new services to those in the existing building that may require the switching off of the existing services, therefore causing disruption to the occupiers.
- Fragmentation of the commissioning process – instead of commissioning the whole building, this may need to be undertaken in zones.

Reflective summary

- Some procurement techniques may prove more beneficial for the client, the contractor or both. In any contractual arrangement it is better to aim for win–win situations.
- The basic control mechanisms for cost, duration and quality of the works that are used on new build contracts should also be used for refurbishment contracts, but they may take slightly different forms.
- Procurement options include: selective or open competitive tendering; two-stage tendering; drawings and specification; negotiation; partnering; cost reimbursement and fee; and design and build.
- There is no procurement method that does not have any risks for the client, the contractor or both. The decision to use a particular method will depend on a number of factors.
- The control mechanisms that contractors use for the management of refurbishment contracts are the same as for new build work.
- However, in refurbishment, because the works can be very fragmented, different contractors could undertake the works in different sequences to produce the same finished product.
- The success of a refurbishment contract is very much linked to the ability, knowledge and conscientiousness of the site team.

Review task

What are the potential risks to the client and the contractor of adopting the following procurement methods for refurbishment work?
(a) Competitive selective tendering
(b) Design and build.

How can access and storage of materials be facilitated in the following scenarios?
(a) Façade retention schemes
(b) Minor refurbishment in an occupied building.

6 Demolition and disposal

After studying this chapter you should have developed an understanding of:

The main drivers and inhibitors affecting the decision to demolish a building
The broad issues which place demolition decisions in context
The main approaches and techniques associated with demolition
The reasons for selecting particular demolition options
The advantages and limitations of different options for demolition

This chapter includes the following sections:

6.1   The demolition decision
6.2   Demolition techniques

- CIRIA (1994): A guide to the management of building refurbishment
- CIRIA R111 (1994): Structural renovation of traditional buildings
- BS 6187 Code of Practice for demolition (BS 6187: 2000 replaces BS 6187: 1982)
- RILEM (2000): Demolition methods and practice; in *Sustainable Raw Materials – Construction and Demolition Waste – State-of-the-Art Report of RILEM TC 165-SRM*
- National Federation of Demolition Contractors: http://www.demolition-nfdc.com/
- Health and Safety Executive: http://www.hse.gov.uk/
- The Institute of Demolition Engineers: http://www.ide.org.uk/

6.1 | The demolition decision

Introduction

- After studying this section you should have developed an appreciation of the factors that are taken into account when arriving at the decision to demolish or dispose of a building.
- You should be aware of the physical, functional and financial issues that must be considered in making an informed judgement.
- In addition, you should have an awareness of the wider environmental issues affecting the decision.

Overview

The issue of building obsolescence is complex, and the reasons for choosing to demolish buildings may often be associated with factors other than physical obsolescence. Buildings will inevitably reach a stage at which their continued use ceases to be appropriate or economic in a given context. Also, they may reach a stage at which their physical performance is such that they are no longer functionally satisfactory for a given use. Such circumstances do not necessarily result in the decision to demolish. Since buildings may pass through several different users, the decision to demolish may be based on a number of factors in addition to those that are related to the physical activity of the occupiers. Several of the factors that affect demolition will be financially driven rather than technological. In the situation where a building ceases to satisfy the requirements of one user, it may pass through several additional users with lesser requirements. As a consequence, the decision to demolish depends on the context of building as well as the physical condition of the structure and fabric. In extreme cases, buildings may become dangerous as a consequence of their poor physical condition, and if left to deteriorate they may inevitably fall into a state which requires demolition. In some situations it may be necessary for the Local Authority to serve a dangerous structures notice on the building owners. Thus it can be seen that the decision to demolish is a complex one and is driven by several factors other than physical condition.

When buildings reach a stage in their life cycle at which decisions regarding refurbishment, future use and the possibility of demolition are to be made, it is important that detailed consideration is given to each of the alternative courses of action. The implications of building lifespan and aspects of life cycle costing will be relevant. The financial implications of complete demolition and rebuilding versus partial demolition, amendment, alteration and extension options must all be considered in the light of whole-life building economics. The process of demolition must not be viewed as the final action in disposing of an obsolete or redundant building. The disposal or redevelopment of the site must also be considered,

PART 4

and as such the act of demolition must be viewed as part of a bigger property development picture. It is not the intention of this text to consider the economics of building development; however, the economic aspects must be considered when making what are often business decisions regarding demolition rather than technical decisions.

Among the numerous factors that contribute to the decisions affecting demolition we must consider the following:

- Building costs in use and other economic factors
- Historic buildings and listed building status
- Physical condition
- Building form
- Sustainability issues.

Building costs in use and other economic factors

The fact that a building is listed does not mean that it cannot be demolished. However, the process of approval is complex.

All buildings are designed with an expectation that at some point they will become obsolete or will have served their intended purpose and will be disposed of. The act of demolition and disposal marks the end point of the building life cycle. Different buildings will inevitably be designed with very different intentions for their expected life cycle, and the potential for redevelopment, alteration and refurbishment will be consequent upon the decisions regarding components and materials at the time of construction. A glance around any town or city will reveal that buildings often survive in functional occupation long beyond their projected lifespans with the aid of major refurbishment and careful maintenance. At various points in the building's life, financial analysis will be required to compare the alternative options for the future of the building. This may result in refurbishment, alternative use or demolition.

In some situations grants may be offered as incentives to retain a building rather than to demolish it.

Historic buildings and listed building status

Many older buildings are of a form that is appropriate for reuse in a modern context, albeit often a different use than the original. However, they will generally require upgrading of the services, fire precautions and interior environment to enable them to be effectively adapted for modern use.

The issue of listed building status was dealt with in some detail in Section 1.5. The implications of listed building status upon the decision to demolish must be carefully considered. Although in principle it is possible to demolish a building that has listed status, the practice is complex and by no means certain. Where buildings are of particular historic or architectural merit it is often the case that an agreement will be reached, resulting in partial demolition with retention of the specific elements of the building that are noted to be of merit, as illustrated in some of the many façade retention schemes of recent years.

Physical condition

Undoubtedly one of the greatest influences upon the decision to demolish will be the physical condition of the building. In extreme cases the Local Authority has powers to serve notices and enforce actions upon owners of derelict and dangerous buildings. The potential to refurbish or alter buildings economically depends upon the existing physical condition of the fabric, and this alone may force the decision to demolish rather than refurbish. As a general rule, the more the condition deteriorates, the more expensive the building will be to refurbish, and the decision to demolish becomes more probable.

Building form

Whilst it may be the case that a building is technically capable of refurbishment, it is often the case also that the layout and form are inappropriate for a desired use. This is illustrated in the examination of office buildings from different periods, for example. A Victorian office building may be in good physical condition, but unable to cater for modern open-plan layouts and the incorporation of cable infrastructure using raised access floors. In such situations, demolition may be the favoured option to allow the construction of a new building of appropriate form.

Sustainability issues

There is a strong drive to improve the sustainability of the built environment. As part of this, there is a growing desire to reuse or recycle buildings rather than entering into the process of demolition. Demolition materials can often be environmentally harmful and the process may result in the release of materials such as asbestos, CFCs and other harmful or environmentally damaging materials and substances. However, the process of demolition can also be managed in a sustainable way by reclaiming and reusing building materials for use on another construction project. Indeed, it is sometimes the case that demolition takes place specifically with a view to reclaiming valuable materials and components that are embedded within a redundant structure. There is a vast and lucrative market for architectural salvage.

PART 4

Reflective summary

- The issue of building obsolescence is complex and the reasons for choosing to demolish buildings may often be associated with factors other than physical obsolescence.
- Buildings will inevitably reach a stage at which their continued use ceases to be appropriate or economic in a given context.
- The decision to demolish depends on the context of building as well as the physical condition of the structure and fabric.
- The following factors need to be investigated before the decision to demolish a building is made: building costs in use and other economic factors; historic buildings and listed building status; physical condition; building form; and sustainability issues.

Review task

Discuss the issues that will affect the decision to demolish a building under the following headings:

(a) Building costs in use and other economic factors
(b) Historic buildings and listed building status
(c) Physical condition
(d) Building form.

6.2 | Demolition techniques

Introduction

- After studying this section you should be aware of the main approaches to the demolition of buildings.
- You should understand the implications of health and safety in selecting an appropriate demolition method.
- In addition, you should be aware of the sequence of operations involved in the process of demolition as applied to larger scale buildings.

Overview

In order to successfully and safely demolish a building or part of a building, it is essential that there is a detailed understanding of the building's structural form and construction. Indeed, the process of demolition requires as much technical knowledge as does the process of construction. Demolition is sometimes undertaken as a sequential process in the reverse order of construction.

In order to gain sufficient information to allow for successful and safe demolition, it is necessary to undergo a detailed data-gathering exercise, which will normally involve the undertaking of a detailed demolition survey. The data gathered as part of the survey will be used to inform the process of selection of an appropriate method of demolition. Available demolition methods are generally defined within two categories: piecemeal demolition and deliberate controlled collapse. Both forms of demolition require the identification of structural form, potential hazards to workers and the public, and limitations that may inhibit successful demolition.

Having undertaken a demolition survey the most appropriate method of work will be selected and various pre-demolition activities, such as serving of appropriate notices, protection of adjacent structures and provision of temporary services, will be put into action. It is perhaps appropriate to consider these activities as pre-demolition issues which are followed by the actual process of demolition.

Pre-demolition issues

The following summarise some of the main issues that must be addressed prior to embarking on the process of demolition.

Demolition survey

The detailed examination and survey of a building that is to be demolished should be considered as essential. Information gathered from this process will be needed to inform the choice of appropriate demolition method and to allow safe and efficient planning and programming of the process. Typically the survey will deal with the following:

- The presence of adjoining/adjacent buildings and structures that might be affected by the process
- The structural form of the building and specific constructional details
- General building condition, with particular emphasis on weak spots
- Identification of requirements for temporary reinforcement or support
- Identification of existing services and utilities
- Presence of hazardous or deleterious materials such as asbestos
- Available access to the site
- Potential dangers to workers and the public.

Having gathered this information, the consultant and contractor will be sufficiently informed to make appropriate decisions regarding the potential options for demolition.

PART 4

Notification requirements

Having made the decision to proceed with the demolition process it will be necessary to deal with a range of notification requirements. These will generally fall into three broad categories as follows:

- Health and safety notices, which may include, for example, notification under the Construction (Design and Management) Regulations 1994, application for licences to allow removal of asbestos and notifications relating to dealing with radioactive materials.
- Local Authority and Highways notifications, including applications relating to noise, intention to dispose of specified wastes, and requests for road closures, sealing of drains and sewers.
- Requests and notifications to utilities companies for isolation/disconnection of services, provision of temporary supplies to site and details of locations of existing services within and around the site.

Health and safety issues

It is clearly recognised that the process of demolition is inherently dangerous. Hence the planning of health and safety management is crucial in planning the overall process of demolition.

The contractor will normally prepare a detailed health and safety plan dealing with the possible hazards likely to be encountered in preparing for demolition and undertaking the actual demolition process.

General considerations relating to health and safety must include requirements for the protection of workers and the public, implications of the presence of overhead electrical cables and buried services, management of the site in terms of access restrictions and safe distances from the demolition area and so on. In addition, consideration of the sequence of demolition must take place to ensure that safe systems of working are assured which take into account, *inter alia*, potential for falling debris, operational areas of mechanical plant, effects of pre-weakening and possible building collapse, and possible use of explosives.

Demolition options

As previously noted, the process of demolition can be undertaken using one of two generic approaches:

- *Piecemeal demolition*: based on demolishing or dismantling the building in a series of interim stages. This process can be effected by hand or with the aid of large mechanical plant or machines.
- *Deliberate controlled collapse*: involving collapsing part or all of the building and clearing debris from the site with excavation plant.

Piecemeal demolition

Although the two alternatives of piecemeal demolition by hand and with machines are identified as separate options, in practice it is often the case that a combination of hand and machine techniques will be employed.

Piecemeal demolition by hand

Piecemeal demolition by hand is generally reserved for buildings of small scale with traditional structural form, and will involve the use of hand tools and assisted access to areas of the building, usually from platforms. This technique is often employed where the potential damage of adjacent elements must be avoided, as in the case of historic buildings, for example. This technique may also be used in situations where the building is to be dismantled and reassembled elsewhere or where part of the building is to be retained. An example of this might be the removal of cladding and roofing from a steel portal frame which is to be re-clad with a new material.

A range of equipment is commonly used to assist with the process of hand demolition, including:

- Breakers
- Cutting equipment and thermal lances
- Hammers, picks etc.
- Lifting equipment.

If using this approach it is important to ensure that suitable working platforms are provided to allow safe access and that any building elements that are cut free are not allowed to fall or swing in an uncontrolled fashion.

Piecemeal demolition by machine

For larger buildings and those that do not warrant special consideration in terms of retrieval of materials and components during demolition, the most common method is piecemeal demolition by machine. It is generally considered that this option is more economical and safer than demolition by hand. It must be acknowledged, however, that the use of mechanical plant in the process of demolition can be restricted by the size and reach of the plant and by environmental considerations associated with the fact that machine-assisted demolition generates more noise and dust than does hand demolition. Notwithstanding this, the accelerated pace of progress that is achievable with mechanical demolition makes it the most popular choice.

A range of mechanical plant is in common use to assist with the demolition process, including the following:

- *Balling machines*: these utilise a large steel ball suspended from a crane, which is normally mounted on a crawler. The suspended ball is dropped onto the building element or swung with the aid of a drag rope to impact against the structure of the building. This is a skilled operation and requires highly competent operatives and well-maintained plant and equipment if it is to be carried out safely.

■ *Impact hammers*: these normally have a track- or wheel-mounted chassis and operate by the use of a heavy-duty pick or pneumatic hammer positioned using an articulated boom. They act to break up large sections of concrete and masonry to allow removal in pieces. This results in high levels of noise generation.

■ *Hydraulic shears and nibblers*: these are fitted to the arms of hydraulic excavators and act to cut through steel members and reinforced concrete.

■ *Pusher arms*: these are generally mounted on tracked vehicles to apply horizontal pushing or pulling force to elements of the structure to induce overturning. The articulated pusher arm is often fitted with a toothed plate or hook to ensure effective application of the force against the building.

Deliberate controlled collapse

Deliberate controlled collapse of part or all of a building can be a highly effective demolition option, but it requires detailed planning and the utmost consideration of safety precautions. It provides a fast and efficient method of building demolition, but is perhaps the most dangerous method of all of the available methods of demolition. The use of explosives is often necessary and it is essential to ensure that appropriate expert advice is taken before considering such an approach. The potential disastrous effects upon adjacent structures must also be considered, as controlled collapse is a process that cannot be stopped once in action.

Collapse of all or part of the building is induced by the removal of key structural elements or by pre-weakening such structural elements before applying controlled lateral forces. These actions may involve the use of controlled explosive charges to weaken or remove elements, often combined with wire-pulling at high level to induce overturning of the structure. The possible sequential failure/collapse of the structure must be understood in detail if this approach is to be safely applied.

Use of explosives

The use of explosives is restricted to certain types of site and structure, although this method has been adopted successfully in many urban sites. The legislation controlling the use of explosives tightly controls the demolition process and restricts use to specified competent persons. In addition, a number of other key safety factors must be borne in mind when adopting this method.

Access to the site and the surrounding area must be restricted to ensure that only authorised persons are allowed in the vicinity of the site. Workers should be issued with a written 'permit to work' to be allowed on the site. Normally this will only be granted when all the necessary safety checks have been made. This process will normally sit within a management regime involving a method statement and a fail-safe 'permit to work' facility.

Emergency services and statutory authorities will also need to be heavily involved in the planning process and they should be consulted at an early stage in project planning.

Pre-weakening

Controlled collapse using pre-weakening does not necessarily involve the use of explosives. However, it still requires careful planning to ensure the safe operation of the demolition process.

In order to promote controlled collapse, key structural elements will be removed or pre-weakened using controlled explosive charges or mechanical plant. In order for this process to be carried out safely, a detailed analysis of the structure must be undertaken to identify which elements can be pre-weakened without inducing premature and uncontrolled failure of the entire structure. The use of temporary supports may be necessary during these operations to prevent the risk of collapse during the pre-weakening operation. During the process of pre-weakening, the removal of strategically located beams, columns and internal wall sections may be combined with cutting of reinforcement steel and creation of holes and voids in the structural elements.

After the structure has been weakened in this manner, lateral forces are generally applied to cause the structure to overturn or collapse. This will be assisted by the use of mechanical plant, including pusher arms and wire pullers. In some instances the use of controlled explosive charges may be adopted to induce the catastrophic failure of the structure.

Reflective summary

- Demolition is sometimes undertaken as a sequential process in the reverse order of construction.
- In order to gain sufficient information to allow for successful and safe demolition it is necessary to undergo a detailed data-gathering exercise, which will normally involve the undertaking of a detailed demolition survey.
- Available demolition methods are generally defined within two categories: piecemeal demolition and deliberate controlled collapse.
- Pre-demolition issues include undertaking a demolition survey, looking into notification requirements, investigating health and safety issues, and considering the most suitable demolition option.
- A range of mechanical plant is in common use to assist with the demolition process.
- The use of explosives is restricted to certain types of site and structure, although this method has been adopted successfully in many urban sites.

Review task

Produce a checklist that could be used for undertaking a pre-demolition survey.

Describe the mechanical systems that are available to help with the demolition process.

Case studies

After studying this chapter you should:

Have developed an appreciation of the complexity of refurbishment projects

Be able to relate the theory of the previous chapters to practical application

Have gained an understanding of the nature of 'unknowns' in refurbishment contracts and how the discovery of these can impact on the cost and duration of refurbishment contracts

Appreciate the factors to take into account when deciding whether to refurbish or demolish an existing building

This chapter contains the following sections:

7.1   Major refurbishment study
7.2   'Decision to demolish' study

7.1 | Major refurbishment study

Introduction

- This section sets out to provide an illustration of the complex nature of refurbishment projects. After studying it, you should have developed an appreciation for the unknown elements that are typical of such contracts.
- You should also be aware of the financial and programming implications of design decisions and the discovery of unforeseen items upon the contract.
- Finally, you should be made more aware of how important a thorough initial survey is to the success of refurbishment contracts.

Overview

This section gives an overview of a major refurbishment contract. The Case study will illustrate some of the problems associated with refurbishment work detailed in earlier chapters, and highlight some of the solutions applied to relatively complex problems. The risks to the client and contractor are discussed and the merits of keeping good records illustrated.

Overview of client requirements and existing building

The existing building was constructed in the early 17th century and is situated in central London. The client and building owner was a major auction house, with an existing property in use at the rear of the proposed development site. Figure 7.1 shows the location of the building to be developed, the owner/client's original building, and the surrounding properties.

The existing building was deemed to be structurally sound in the front area as shown on the diagram, but the rear of the building was showing signs of significant structural problems (Figure 7.2). Cracks in the parapet at roof level were evidence that there was structural movement. These were monitored to see if they were recent or longstanding, and whether movement was still ongoing.

The client brief to the designer was to allow for the following in the refurbished building:

- Basement to be used for storage
- Ground floor: provide one new sales room and a prestigious entrance
- First floor: provide two new sales rooms
- Second–fifth floors: provide flexible use office space
- New male and female toilets on each floor
- Repairs and renewal to front elevation in keeping with surrounding area
- Provision of two goods lifts and one passenger lift
- New services installation.

Figure 7.1
Proposed development site.

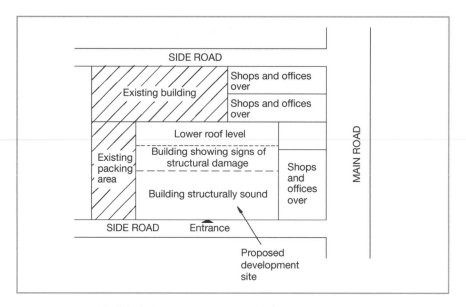

Figure 7.2
A crack in a parapet.

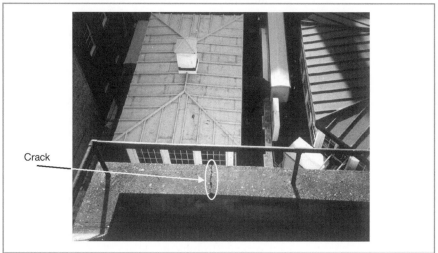

Planned operations

The planned operations for the scheme were as follows:

1. Remove some of the existing features of the building and refurbish away from the site for replacement after major works were complete.
2. Demolish the rear half of the building and expose existing steelwork, and demolish the entire roof (Figure 7.3).
3. Undertake bored *in situ* concrete piling in the area shown (Figure 7.4) in order for a new pile cap to be constructed.
4. Erect a new steel frame (Figure 7.4) on the pile cap and connect this new frame to existing steelwork at the rear of the building up to first floor level.

Figure 7.3
Demolition parameters.

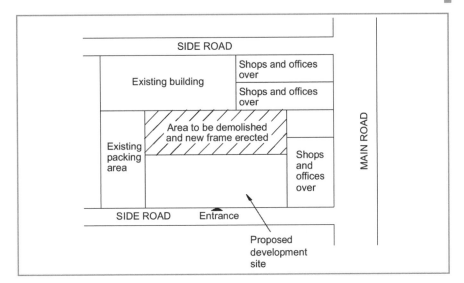

Figure 7.4
Extent of new steel frame.

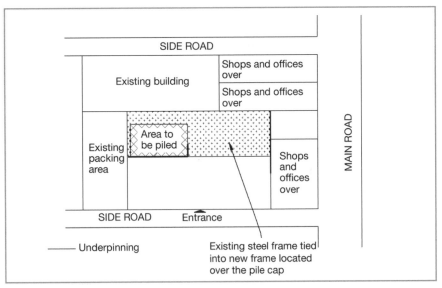

Continue with the erection of the steel frame to the entire rear half of the building up to roof level.

5. Fix steel decking to the new steel frame.
6. Concrete-in steel beams and columns.
7. Construct brickwork cladding to the rear of the building.
8. Repair and renovate the front elevation of the building.
9. Build three lift shafts and install lifts.
10. Install modern mechanical, electrical and plumbing systems.
11. Build roof structure and coverings.
12. Provide finishes to upper floors and basements.
13. Provide high-quality finishes to sale room and reception areas.

Extra activities required, problems and site management issues

Access to the site

Access to the site was severely restricted, with the road where the site entrance was located being very narrow. This meant that the size of lorry that could deliver to the site was relatively small and large materials orders had to be staggered. This had a cost implication, because suppliers required additional payment for the extra deliveries. It also meant that many deliveries had to occur out of traditional working hours so as not to inconvenience the occupiers of the surrounding buildings.

It was planned that a chute would be formed through the pavement, and materials would be lowered into the basement. Holes were then to be broken through the existing floor slab on every floor in order for a materials hoist to be installed. Materials would be loaded in the basement and lifted to each floor using the hoist.

Figure 7.5 shows the position of the chute (e) and the location of the hoist (a).

The hole was broken out and steel Universal Beam sections fixed to the sides of the opening using bolts. The gap between the steel and the floor was made good using concrete. The intention was that after the back of the building was rebuilt, the hoist would be moved to the position of the new passenger lift shown in Figure 7.5, location (b).

> It is common practice to install hoists in the location of new lift shafts.

The hole in the existing floor would then be made good by placing steel decking in the void, shot firing it to the steel trimmers, placing anti-cracking mesh on the decking and concreting to the original floor level.

Figure 7.5
Access for materials.

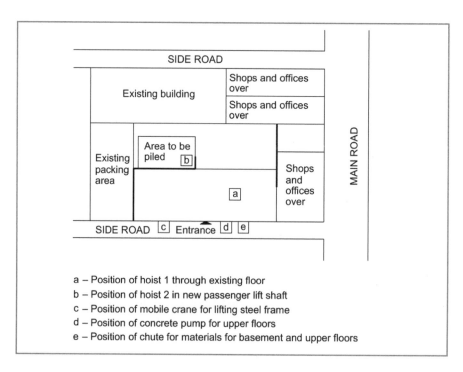

a – Position of hoist 1 through existing floor
b – Position of hoist 2 in new passenger lift shaft
c – Position of mobile crane for lifting steel frame
d – Position of concrete pump for upper floors
e – Position of chute for materials for basement and upper floors

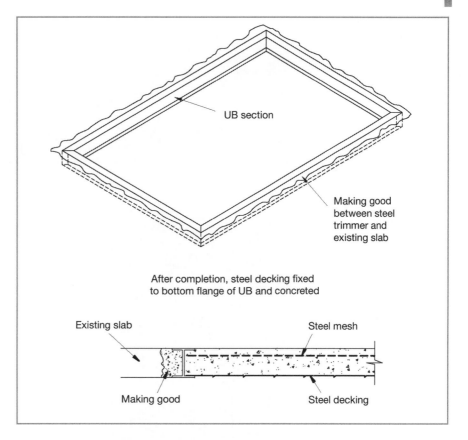

UB section

Making good between steel trimmer and existing slab

After completion, steel decking fixed to bottom flange of UB and concreted

Existing slab

Steel mesh

Making good

Steel decking

Figure 7.6 shows how the void was created and then made good. Figure 7.7 shows the floor when it was being broken out (left) and the hoist supporting scaffold at the roof level (right).

During the breaking out of the floor activity, a problem arose. The existing floors were not what they seemed! They were constructed using a framework of steel beams, and alternate floor panels were concrete and then a mixture of straw and plaster. These had stood the test of time and showed no evidence of excessive cracking or deflection, but would not comply with current Building Regulations. All of these panels had to be located, broken out and replaced with new concrete. This was a significant extra and there were cost and progress implications.

The new steel frame that was to be constructed at the rear of the building had to be lifted into place, and it was not possible to use a tower crane due to objections from the surrounding building owners about a crane going over the top of their buildings. It was therefore planned that a mobile crane be used to enable erection of the steel frame. It was planned that the location of the mobile crane would be as shown in Figure 7.5, position (c). In order to facilitate this, closure of the road was required. It was planned to erect the frame over four consecutive weekends, and closures were applied and paid for. However, due to delays these dates were missed, and the result was that the steel frame was carried into posi-

PART 4

Figure 7.7
Breaking out the floor (left)
and hoist supporting
scaffold (right).

Figure 7.8
Demolition rubble from the
basement reaches first floor
level.

tion by the steel erectors. The erection of the frame took a great deal longer and therefore cost more to erect.

Once the steel frame was erected and steel decking fixed in place, the floors needed concreting. It was planned that a vehicular concrete pump would be used, but on the day of the first pour, the pump was wheel-clamped due to it partially blocking the road. A smaller mobile pump had to be used that had to be stored within the site boundary when not in use and pulled into position every time it was required. The pump was slower than the vehicular pump and therefore the concreting activity took longer than was anticipated. The pump was located when required for concreting at location (d) on Figure 7.5.

It was impossible to get an excavating machine into the building and therefore all demolition waste had to be loaded into and carted out of the building in wheelbarrows.

Figure 7.8 was taken at first floor level from the remaining first floor slab. The rubbish that can be seen is demolition rubble that has built up from the basement – approximately 10 m!

Piling

The initial scheme proposed that nine new piles would be required. However, when the piling activity started two old wells were discovered. These had to be filled with concrete and an additional two piles were required in order to stabilise the pile cap.

Underpinning

During excavation work at the rear of the building it was discovered that the building adjacent to the site and the back end of the building that was to remain in place required underpinning. This was due to there being no foundations under these areas. Mass concrete underpinning was carried out in these areas,

Figure 7.9
Extent of underpinning works.

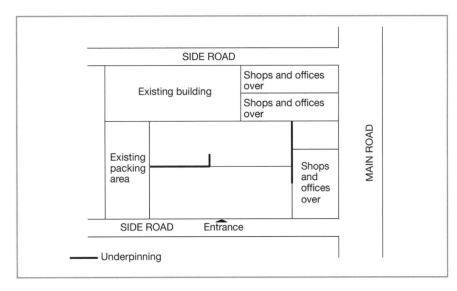

Figure 7.10
Proposed gridlines.

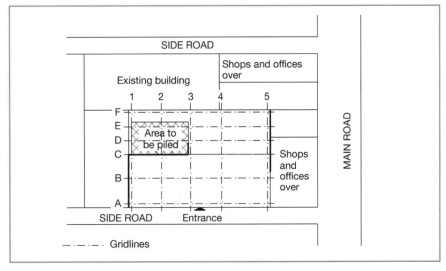

and this caused an increase in cost and delayed the project. The area that required underpinning is shown in Figure 7.9.

Setting out

This was a refurbishment contract, and part of the existing building was remaining, but a new steel frame was being attached to an old structure. The tolerance for the setting out of new steelwork is 3 mm, and therefore the setting out needed to be very accurate. Also, the new interior design of the remaining half of the building required a perfectly square grid system. The setting out line positions are shown in Figure 7.10.

However, it was impossible to sight from the front of the building to the rear with a theodolite due to the existing structure. The only option was to break holes

Figure 7.11
Establishing gridlines using a plumbline.

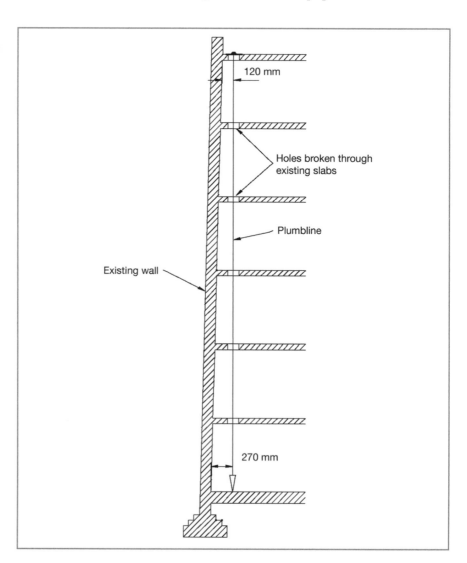

through the floors and drop down a plumbline from the roof level to the basement, as shown in Figure 7.11.

When this was carried out it was found that the external wall of the building was out of plumb by 150 mm. The interior design of the existing building then had to be revised significantly to account for this and some additional steelwork was required in order to allow the new frame to attach to the old. There were significant cost and delay implications because of this.

It is fairly common for existing building walls to be out of plumb and out of square.

Additional steelwork

Figure 7.12 shows the remaining half of the building after demolition, with floors broken out, some of the new steelwork fixed to the pile cap and existing steelwork running along gridline C and between gridlines 1 and 3.

It was proposed that in the area where piling had been undertaken (enclosed by gridlines CDEF/123), a new steel frame would be constructed the whole height of the building. In the adjacent areas (enclosed by gridlines CDEF/345), new steel would be connected to existing steel goal posts that had been exposed during demolition at first floor level. However, when the original steel was exposed, what was not anticipated was that the goal posts would not be connected to the steel work in the remaining structure. Additional steelwork therefore needed to be designed and fixed to the goal posts and the steelwork in the existing half of the building to provide stability. Figure 7.13 shows this additional steelwork.

Basement slabs

While the contractor was undertaking an initial survey of the building, it was identified that the floor-to-ceiling height was not as shown on the drawings. After installation of the new services the floor-to-ceiling height would be unacceptable. Therefore an instruction was given to the contractor to lower the existing basement floor levels by 300 mm. The existing floors were to be broken out and soil excavated down a further 300 mm. Traditional slabs were then to be installed comprising a layer of hardcore, blinding, DPM, steel mesh and concrete. Due to problems with access, the breaking out of the slabs and excavation had to be carried out by hand. This again was very costly and took a great deal of time. This particular activity did not, however, cause a significant delay, as no work was planned in the basement at this stage of the contract.

Works to the front elevation of the building

The specification for the front elevation of the building works was simply to clean and undertake minor repairs. However, once the building was cleaned, the balconies were different colours. This was due to the fact that during the Second

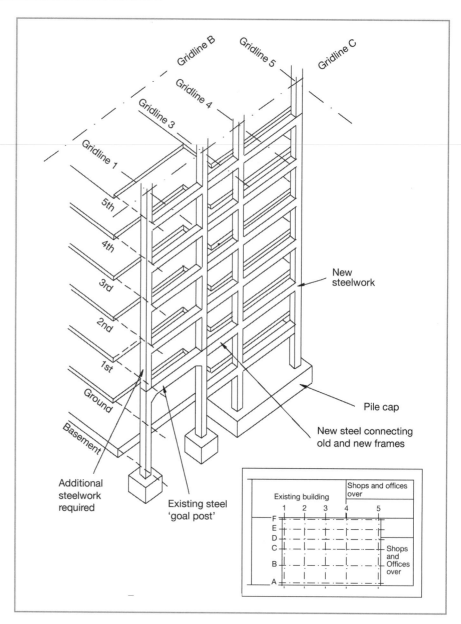

Figure 7.12
Steelwork details.

World War some of the balconies had been demolished because of bomb damage and replaced with balconies made from precast concrete, as opposed to the originals which were carved out of stone. The cleaned façade actually looked worse than before. The designer therefore specified the need for extensive painting of all of the balconies. Figure 7.14 shows the balconies after cleaning and then after repainting.

Figure 7.13
Additional steelwork required.

Figure 7.14
Cleaning and repainting the balconies.

Interior design

The interior design for the sales rooms required perfectly square rooms. However, after setting out in all three sales rooms, it became apparent that this was not the case. Therefore to make the room perfectly square, metal studding was fixed to metal sole and soffit plates and plywood sheets fixed to the studs to form partitions that act as false walls.

Figure 7.15 illustrates how this was achieved and the extent to which the existing room was out of square.

Figure 7.16 shows how important it was that the room was made to be square: if this had not been done, there would have been unsightly 'cuts' around the perimeter of the ceilings.

The suspended ceiling is a 600×600 mm modular concealed grid system with plaster tiles that have then been painted white.

PART 4

Figure 7.15
Plan layout of room and jumbo studwork layout.

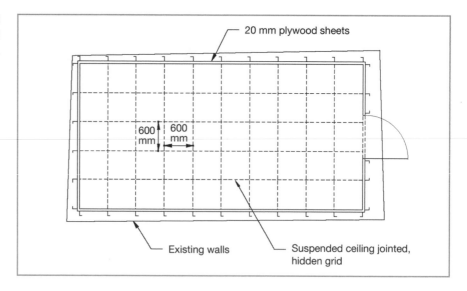

20 mm plywood sheets

600 mm | 600 mm

Existing walls

Suspended ceiling jointed, hidden grid

Figure 7.16
The squared-off ceiling.

Floor-to-ceiling heights in older buildings tend to be greater than required in new buildings, and this allows for the inclusion of raised access floors and suspended ceilings.

The floor is a shallow batten raised access floor finished with timber strips that can be removed to gain access to the services underneath, which are mainly electrical wiring. The restricted floor-to-ceiling height in the original building limited the floor void depth to 50 mm.

The majority of the services are enclosed behind the partitions and within the suspended ceiling.

An interesting feature of this room is that the wall covering is carpet. The reason for the choice of this material as a wall covering is that pictures are to be hung and changed regularly. Using nails will damage virtually every alternative covering, but when nails are removed from the carpet no sign of them is evident.

Figure 7.17
The main staircase.

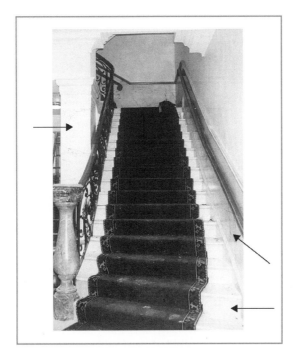

In the ground floor of the existing building, marble had been used extensively as a finish. The client wished to retain this marble and it was protected before any construction work began. However, when the protection was removed it looked very shoddy when compared with the new finishes. Initially the client tried to blame the contractor for damaging the marble, but the contractor had undertaken a very detailed schedule of conditions before work commenced and could therefore prove that the damage was not caused by them.

The cost of replacing the original marble, together with the very long lead-in time required to procure the material, meant that an alternative was required. It was decided that all marble would be replaced with plywood that would then be painted to look like marble. This required the skills of very specialist painters, who required a large sum of money in order to undertake the works.

Figure 7.17 shows the main staircase when complete. All the areas highlighted have been painted to look like marble columns: the staircase post, the wall beneath the dado rail, and the stair treads and risers.

Accommodation

The site was very restricted and work was planned in every area of the site. Therefore the use of portable buildings for accommodation was impossible. The site was in central London and hence the only office space available in the area was extremely expensive. It was therefore decided that the existing parts of the building would be used for accommodation. Figure 7.18 shows the location of offices over the contract duration.

PART 4

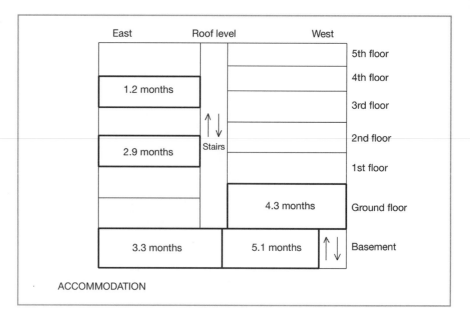

Initially accommodation was located on the fourth floor because extensive works were being undertaken on the lower floors and the hoist was located in the other half of the building. On the second floor was a planned control room that was not going to be fitted out until late in the contract. Once the walls were built the accommodation moved to this area for nine months. When it was time to fit out the control room the accommodation needed to move, and went into the basement for the next three months. Then it was time to undertake internal finishes to the basement, so the accommodation needed to move and went to the first floor for the next three months. Once the basement works were complete, the accommodation moved back from the ground floor to the basement for the final month of the contract. Every time the accommodation moved it would take a gang of labourers three to four days to move everything and caused serious disruption. There was a cost implication and other activities had to stop to facilitate the move.

Cost and duration implications

All of the above required additional funding to be made available and caused serious disruption to the programme. The completion date could not be changed as the client had planned and been advertising auctions three days after the original planned completion date.

It is extremely rare for new build contracts to go over tender price by 17 per cent, but much more common in refurbishment contracts.

The original tender value was £5,200,000, which included the planned profit to the contractor of 2 per cent, £104,000. The actual final costs of the work were £6,100,000, an increase of £900,000 (17 per cent). There was no extension of time given, but a £500,000 acceleration package was negotiated by the contractor. Due to the competence of the contractor's management team, the actual profit to the contractor was £1,330,000, which is 25.5 per cent of the original contract value. This was achieved by careful management and the keeping of good records.

Reflective summary

- It can be seen from the details of the Case study that the points made about the need for detailed surveys before the design of refurbishment work is undertaken were significant.
- It can also be seen that the points made regarding successful management of refurbishment work were all adopted by the contractor and resulted in increased profits.
- The increase in costs because of unknowns is clearly illustrated.
- This building was a refurbishment contract, but it was also a conversion because the building had planned different usage. There were elements of the building that were renovated and/or repaired. Elements were renewed and a complete new services installation installed; hence retrofitting was required. Therefore this contract was a combination of all the terms that are sometimes mistakenly applied to refurbishment contracts.

Review task

What could have been done to avoid all of the extras and variations that occurred on this contract? Deal with each point individually and refer back to Figure 1.5 for guidance as to what should be done when undertaking a feasibility study.

7.2 | 'Decision to demolish' study

Introduction

- After studying this section you should have gained an appreciation of the factors that are taken into account when reaching decisions regarding the potential refurbishment or demolition of an existing building.
- You should be aware of the issues associated with building functionality, spatial configuration and flexibility that inform the decision to demolish rather than rebuild.
- In addition you should understand the various issues associated with long-term maintenance and operating costs together with the increasing importance of carbon management and reduction in the context of property assets.

Overview

This section gives a summary of a demolition project associated with a six-storey educational building. It sets out the salient points that are taken into account when considering the demolition versus refurbishment decision and the relative weight of each of the factors that are considered. In addition, the benefits and disadvantages of each of the options are summarised along with the key drivers that affect the corporate decision-making process. The technical, functional and economic considerations are summarised and key stakeholders in the decision-making process are identified.

Existing building overview

The building that forms the basis of this Case study is a six-storey (plus basement) educational building that was constructed in the mid-1960s. The structural form is very traditional and is as follows:

Structure and fabric

Structural frame	Steel frame, encased in concrete for fire protection.
Floor and roof structure	Concrete beam and block form.
Exterior cladding	Bands of brick cladding (uninsulated).
Windows	Bands of galvanised steel casement windows.
Roof covering and detail	Asphalt flat roof with brick parapet, capped with reinforced concrete copings. Fitted with retrofit insulation in 1980s.
Access	Central lift/stair core with two lifts and two stairs.

Services

Heating	Gas-fired modular boilers, retrofitted in late-1990s. Cast iron radiators on low-pressure hot water circuits (pipework uninsulated).
Air-conditioning	Comfort cooling to ground-floor lecture theatre retrofitted in late-1990s.
Lighting	Fluorescent tube fittings to most areas, upgraded for VDU use on an *ad hoc* basis.
Controls and BMS	Limited central control of heating in two simple zones. Small number of thermostatic radiator control valves on *ad hoc* basis.
Access	Two electric traction lifts with simplex control (i.e. no intelligent control of the lifts as a pair).

Figure 7.19 shows the building, which clearly looks dated and to some extent dilapidated. The general arrangement of a sample floor of the building is illustrated in Figure 7.20. The internal layout is typical of commercial and educational buildings of the period, featuring a central access core in the middle of the building. This core houses the two electric traction lifts and two stairs, separated by a central spine wall for fire escape purposes. The usable space is positioned around the perimeter of the building, with a central access corridor around the access

Figure 7.19
Photo of Clarence Street building.

Figure 7.20
General arrangement (floor layout).

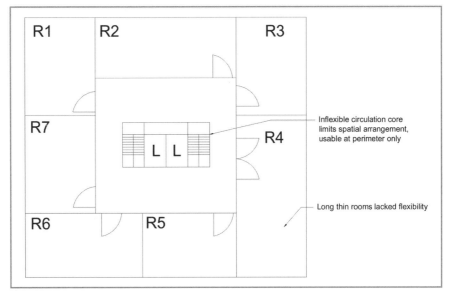

core. All of the usable rooms within the building are accessed from the central circulation space.

Summary of client requirements

The building that is the basis of this Case study forms part of a large portfolio of educational buildings serving a city-based University. As changes to educational delivery have evolved, so the needs of educational buildings have evolved and the University estate is subject to continual review and enhancement to serve these changing needs. In order to inform the decisions regarding individual buildings, an overall property strategy has been developed that supports the operational needs of the University going forward. As part of this strategy, facilities for future operational needs are defined and scoped. In addition, the suitability of the existing facilities is constantly reviewed and assessed.

In the case of the chosen building there was a recognised need to attend to the increasing level of deterioration of the building. Although a planned maintenance programme was in place the building was suffering from increasing levels of failure of components. Despite increasing levels of reactive and planned maintenance expenditure, the level of disrepair was increasing. The costs of arresting this process were disproportionate to the functional value of the building.

This process of declining performance over time is covered earlier within this book and the point at which a decision to embark on major refurbishment or to demolish and rebuild had been reached. Given this scenario and the broader strategic intentions for the development of the estate into the future, an option appraisal was considered necessary. This reviewed the possible ways in which the clients' property strategy could be achieved for this building. The intended strategy was to create a flexible, modern facility that could cater for high levels of information and communication technology. In addition, the imperative to provide a sustainable estate with reduced carbon emissions demanded an energy-efficient, environmentally managed facility. There was also recognition that the nature of the use of buildings is changing constantly and this drives for the creation of a building with sufficient flexibility of form to cope with changing future needs. As a result, the client needed to consider carefully the extent to which the existing building could achieve these requirements, either in its existing form, or following refurbishment and adaptation. The relative cost of refurbishment versus demolition and redevelopment was a key factor in the decision process.

Basis of demolition decision

In the selected case, the decision to demolish and rebuild was taken as a result of a combination of physical, functional and economic factors. The first thing that was considered was the degree to which the existing building was able to fulfil the defined client requirements in physical and functional terms.

Perhaps the easiest way to consider this process is as a matter of supply and demand. The current and future needs of the organisation represent the demand, which must be articulated in terms of space, functionality, sustainability and costs in use. The supply side of the equation is represented by existing building stock, either as it exists or through a process of refurbishment, or is delivered by the creation of new buildings. The choices that are presented in considering how to satisfy this simple supply–demand equation essentially represent the refurbish versus demolish decision process.

The factors that feature on the demand side of the equation are summarised in very simple terms in Figure 7.21, together with the existing building's capacity to meet the deliver side of the equation, either as existing (E) or following refur-

Figure 7.21
Supply and demand
matrix.

Demand		Supply		
		Good	**Neutral**	**Poor**
Space	Quantity	N	ER	
	Configuration	N		E R
	Flexibility	N		E R
Functional capability	Capacity to support intended use	N	R	E
	Suitability of space and facilities	N	R	E
	Fitness for purpose	N	R	E
	Technological capability	N	R	E
Performance in use	Energy costs	N	R	E
	Environmental comfort	R N		E
	Maintenance costs	N	R	E
	Carbon management	N	R	E
	Freedom from defects	N	R	E
Strategic factors	Balance sheet asset value	N	E R	
	Corporate image	N	R	E
	Fit with future estate strategy	N		E R

bishment (R) and compared with a possible new building (N). The resulting matrix clearly indicates the gap between the ability of the existing building to cater for the organisation's needs and the capability of a new, purpose-built facility.

The matrix illustrated in Figure 7.21 indicates clearly that the existing building, even if refurbished, would still fall short of the performance level of a new building. There are several key reasons for this as follows.

The existing floor layout with the central access core limits flexibility in spatial layout, and all rooms would need to be accessed from a central circulation space. As such, the building operates as a series of concentric zones as indicated in Figure 7.20. The configuration of the usable space would always be restricting upon the ability to remodel the outer zone, which is limited to long, thin sections that do not suit a flexible, loose fit arrangement. In addition, the existing building did not maximise the utility of the site that it occupied; a new building could exploit the site far more effectively and efficiently.

The extent to which the performance of the building fabric could be upgraded was limited by the technical solutions available, such as overcladding. The ability to alter fenestration to mitigate the effects of heat loss was limited due to the existing form. Hence, even following refurbishment the performance would fall short of that of an equivalent new building.

The level of deterioration and obsolescence of the building and its components resulted in the only viable refurbishment option being a major, costly, remodelling. The cost of this relative to a new build option was not justifiable. Essentially, the building would need to be stripped back to the structural frame and refurbished with all new services, fabric, fixtures and fittings if an acceptable performance level was to be achieved. The cost of such an approach would not be economical.

Hence the decision was taken to demolish the existing building allowing a new, sustainable, building to be constructed that has the added benefit of enhanced corporate image for the client. The process of demolition is captured in the sequence of photographs given in Figure 7.22.

Figure 7.22
Demolition photos.

Reflective summary

■ The physical condition of the building was a key element in driving the decision to demolish.
■ Even though the building was technically capable of refurbishment, it would not have been able to fully satisfy the users' current and future requirements.
■ The cost implications of demolition versus substantial refurbishment are less onerous than it may first appear.

Index

air conditioning   165
asbestos   10, 37, 195, 203

basement   122
  ceiling   130
  tanking   124
brick
  cladding   84, 149
building
  appraisal   17
  condition   48, 65, 204
  control   27
  obsolescence   14, 199
  services   165

carbon management   223, 227
carbon reduction   18, 23, 24
carbonation of concrete   86
categories of listing   29
cavity
  tray   87, 103
cavity walls   173
  wall tie failures   78
CDM regulations   37
CFCs   201
change of use   17, 47, 52
cladding   103, 133
cleaning buildings   154, 192, 218
condensation   61, 66, 81, 84, 90,
    101, 103
conservation of energy   15, 23
conversion   6
costs   8, 9, 193, 222
cracking   67, 72, 73, 117, 132, 147,
    152, 174
cracking, monitoring   68
curtain walling   40, 83, 150

dampness   101
  penetrating   101, 102
  remedies   169
  rising   102, 169
damp-proof courses   102, 127
  remedial   127
defects
  analysis   67
  diagnosis   57
  origins   61
deleterious materials   15, 203
demolition   198
  criteria   199
  methods   202
  procedures   204
design   13, 21, 40, 184
differential movement   132, 137
disproportionate collapse   226
drainage   48, 89, 108, 125
dry rot   95
  remedial treatment   176

energy conservation   15, 23, 24
energy consumption   18, 25
environmental impact   8, 18, 20,
    39, 201
environmental performance   23, 53
environmental preference method
    19, 20
external
  cladding   39, 146
  insulation   159

façade retention   132, 144
fire
  historic buildings   195
  protection   141

fire – *continued*
  safety legislation   53
flat roofs
  coverings   89
  defects   87
floors   92, 98, 125
foundations   17, 44, 152, 183,
    190
framed buildings   63, 73, 79, 86
fungal attack   95

gutters   41, 89

health and safety   35, 39, 184,
    194, 195
heave   71
historic buildings   28, 200, 205

insect infestation   94, 177
insulation   23
interstitial condensation   103

joints, movement   93

life cycle of buildings   6
listed buildings   29

maintenance   46, 47, 51
  cyclical   48
  forms of   48
  planned   48
  programming   49
  reactive   48
masonry repairs   173
movement
  differential   132, 137
  joints   92

obsolescence   14
overcladding   146–55
overroofing   156–64

Part L   22, 147, 167
Part M   27, 31, 185
partial demolition   199
penetrating damp   101, 102
piled underpinning   120
pitched roofs   156, 160

raking shores   135
refurbishment   5, 16
regulations   36
remedial treatment
  dry rot   176
  insect infestation   177
  wet rot   177
renovation   6, 185
resin anchors   142
retrofit   6
retrofit services   165–8
rising dampness   102, 169
roofs
  flat   87, 89
  overroofing   156

scaffolding   36, 135
security   51, 54
services   165, 167
sound insulation   149
sulphate attack   81
sustainability   5, 18, 21, 24, 201

tanking, basements   127
temporary support   132, 143
timber decay   78, 101
town planning   29, 40, 184

underpinning   117

wall tie failures   78
wall tie repairs   173
wet rot   177
whole life cycle   5, 219